A BASIC COURSE IN

COMPLEX VARIABLES

by

David C. Kay*

*A former teacher of college mathematics

A BASIC COURSE IN COMPLEX VARIABLES

iUniverse books may be ordered through booksellers or by contacting:

iUniverse
1663 Liberty Drive
Bloomington, IN 47403
www.iuniverse.com
1-800-Authors (1-800-288-4677)

ISBN: 978-1-4917-4265-5 (sc)
ISBN: 978-1-4917-4266-2 (e)

Library of Congress Control Number: 2014913519

Printed in the United States of America.

iUniverse rev. date: 01/13/2015

To Katie

CONTENTS

PREFACE

This version of complex variables provides a short but comprehensive coverage for undergraduates that features complete explanations of all concepts, key examples for enhanced understanding, and carefully-crafted problems. It is intended for the typical mathematics major and for students in near math-major status enrolled in areas of science and engineering. The presentation develops the subject from quite elementary concepts of algebra, gradually advances to elementary concepts of calculus, and evolves into more sophisticated ideas. The reader is expected to have a working knowledge of basic calculus, and as the material progresses, to have or acquire a knowledge of multivariable calculus.

No serious treatment of the subject can remain entirely elementary however; many advanced topics in mathematics are unavoidable. But instead of being merely brushed aside as "beyond the scope" as in many treatments, an effort is made to deal with them here. Many such topics occur as guided problems independent of the main text. These topics may be assigned as homework or covered in class, providing flexibility for the instructor. All other topics actually covered in the text are introduced and motivated in terms that a typical mathematics major can follow. Certain topological concepts are necessary for Cauchy's theorem, for example, and here the concepts are carefully explained, accompanied by illustrative diagrams. Another example of a necessary advanced topic is that of uniform convergence. It enters the discussion intuitively when Taylor's theorem and infinite series appear in Chapter 6, and is fully developed later in Chapter 7, preceding the development of Laurent series in Chapter 8. In Chapter 9, a collection of traditional topics appears, including a discussion of mappings, fixed points, the Julia and Mandelbrot sets, analytic continuation, and infinite products.

The contrast between real variables and complex variables is a central theme. The student learns how the concept of differentiation in complex variables leads to unexpected consequences. Unusual ideas which typical undergraduates should find interesting are emphasized, eventually revealing the exquisite power of complex variables to solve problems in real analysis. One of the features of the book is a relatively simple proof of the Cauchy-Goursat theorem, which makes it unnecessary to cover extraneous preparation material, such as proofs involving Green's theorem (requiring continuity of the partial derivatives).

This book is highly recommended for mathematics departments offering a traditional undergraduate course in complex variables. It might also be used as a supplementary text for a course in foundations (at least for those courses that emphasize real analysis) since many standard topics in that area are covered. The student can witness first-hand the purpose for such topics, a feature often lacking in traditional foundations courses. In all, this book offers a refreshing treatment that can serve a variety of needs, either as a textbook, as a tutorial to accompany other textbooks, or even as a self-help guide for the student.

My credentials as an author of a book in this area include the following: I was a mathematics college professor for over 30 years and taught in the mathematics departments at the Universities of Oklahoma and North Carolina (Asheville), where I served as chairman for nine years, retiring in 1997. I am the author of several other mathematics textbooks. (See back cover for more information.)

—David Kay
Athens, Georgia

PREFACE FOR STUDENTS

What are *complex variables*? One could say that they are *arbitrary complex numbers*. Such "numbers" were no doubt first introduced to you in elementary algebra as quantities having the form $a + bi$ (like $2 + 3i$ and $4 + 5i$) where $i = \sqrt{-1}$. In Chapter 1 we show how to remove some of the mystery you might have experienced in working with such numbers. By developing a number system in which ordinary arithmetic is meaningful, and one that includes the complex numbers $a + bi$, it becomes clear that the ordinary rules of algebra are mathematically valid for complex numbers, and a concrete starting point for the study of complex variables emerges.

The next idea is the concept of a function. In Chapter 2 we develop logical definitions for such functions as $\sin z$ and $\ln z$, where z is a complex number. In fact, all the basic functions of calculus are extended to include complex variables, and we show how to make specific calculations using them. There are lots of illustrations to help you understand all these ideas. Many carefully crafted examples focus on the concepts as you progress through the book.

In this book, you will need to have a working knowledge of not only algebra, but of calculus. Multivariable calculus becomes necessary in order to appreciate the material in Chapter 3. The derivative of a complex function is introduced—the same one you learned in calculus. We find out, for example, that the derivative of $\sin z$ is $\cos z$, and the derivative of $\ln z$ is $1/z$, just like ordinary calculus. Fortunately, you will not have to learn new formulas. Even the methods for differentiation are the same, such as the derivative of a product and quotient, and the chain rule.

So if all the basic theory is the same, why study complex variables? The fun does not really begin until the concept of the integral is introduced in Chapters 4 and 5. One result of this development is that if the *first* derivative of a function exists, then its *second* derivative also exists, unlike ordinary calculus. This result is a consequence of a famous theorem in complex variables due to A. Cauchy (1789–1857). Cauchy's theorem states that if you integrate a differentiable function about a closed curve, you always get zero. It is surprising how much can be gleaned from this one simple (and seemingly trivial) result. It is the cornerstone of complex variables, however. The ultimate consequence is that complex variables has much greater power to solve problems than ordinary calculus.

Chapters 6, 7, and 8 take us deeper into the theory and reveal wonderful methods for solving difficult problems, both *pure* (mathematically interesting but of no apparent use) and *applied* (of practical value, sometimes of lasting importance). The usual topics in this area were once considered graduate level material, but this book makes these topics accessible to a typical (but serious) undergraduate.

Finally, Chapter 9 presents some really unusual concepts, having a wide range of interest and application. Among them is a study of mappings induced by complex functions. Included in this chapter is the fascinating area of infinite products. (You may recall the Wallis products briefly discussed in calculus.) Here unique formulas for the infinite product expansions of $\sin z$ and $\cos z$ appear. In conclusion, this book is a relatively short account of unique results that have been developed by great mathematicians in the past. Their work and the results they discovered have made it possible for us to enjoy the area of complex variables in the best possible manner, often with very little effort. Let our journey begin.

1

COMPLEX NUMBERS
AND COMMON NOTIONS

WORKING WITH IMAGINARY NUMBERS

You no doubt once solved quadratic equations having solutions that were *imaginary*, or *complex*. For example, consider $x^2 - 4x + 13 = 0$. If you use either the method of completing-the-square or the quadratic formula, you come up with an answer of the form $x = 2 \pm 3i$, where $i = \sqrt{-1}$. When you were first introduced to this, you were probably told (perhaps mysteriously to you) that *i is an imaginary number whose square equals* -1, *and that it should be treated just like an ordinary "algebraic quantity"*. For example,

$$3i + 2i = 5i, \qquad 3i \cdot 2i = 6i^2 = -6, \qquad 4i(3 + 2i) = 12i + 8i^2 = -8 + 12i$$

and so forth. That is, all the ordinary rules of algebra are valid. (One might ask, "how do we know that for sure?")

Since we are going to be calculating with these kinds of numbers exclusively, it is good to review how they are supposed to work in elementary algebra. The steps in solving the above equation by completing-the-square are, as follows:

$$
\begin{aligned}
x^2 - 4x + 13 &= 0 \\
x^2 - 4x \quad\;\; &= -13 \\
x^2 - 4x + 4 \;\; &= -13 + 4 \\
(x - 2)^2 &= -9 \\
x - 2 &= \pm 3i \quad \text{(where } i = \sqrt{-1} \text{)} \\
x &= 2 \pm 3i
\end{aligned}
$$

To check this result, one substitutes $2 \pm 3i$ for x, making use of the fundamental property $i^2 = -1$. Thus (using just the one root $x = 2 + 3i$),

$$
\begin{aligned}
(2 + 3i)^2 - 4(2 + 3i) + 13 &= 0 \\
(2^2 + 2 \cdot 6i + 9i^2) - 8 - 12i + 13 &= 0 \\
4 + 12i - 9 - 8 - 12i + 13 &= 0 \\
4 - 9 - 8 + 13 &= 0 \\
0 &= 0
\end{aligned}
$$

1

In general, and for future reference, we can list some basic identities involving this imaginary number i, all obtained by "ordinary algebra."

Powers of i: $\quad i^2 = -1, \quad i^3 = -i, \quad i^4 = 1, \quad i^5 = i, \quad \cdots$

Powers of $x + iy$: $\ (x + iy)^2 = x^2 - y^2 + 2xyi$

$\qquad\qquad\qquad\quad (x + iy)^3 = x^3 - 3xy^2 + (3x^2y - y^3)i$

$\qquad\qquad\qquad\quad (x + iy)^4 = x^4 - 6x^2y^2 + y^4 + (4x^3y - 4xy^3)i$

$\qquad\qquad\qquad\quad (x + iy)^5 = x^5 - 10x^3y^2 + 5xy^4 + (5x^4y - 10x^2y^3 + y^5)i$

TABLE 1

ALGEBRA AND IMAGINARY NUMBERS

Were you ever bothered by being told that the "number" $i = \sqrt{-1}$ was to be treated as an ordinary algebraic quantity (like x), and to use the ordinary rules of algebra to work with it? For example, to find the product $(2 + 3i) \cdot (5 - i)$ you were instructed to just use the distributive law as in ordinary algebra. Never mind that $\sqrt{-1}$ exists only in your imagination; certainly it does not exist in the "real world" we live in, and you will certainly not find it on an ordinary pocket calculator. Even assuming that i can be given some type of meaning, how can we be certain that the laws of algebra will not lead to a contradiction of some sort? Any such contradiction would be quite devastating.

Incidentally, the history of i is an interesting story of an evolving mathematical concept that engaged the minds of many great mathematicians of the past. They wrestled with these same questions. Leonhard Euler (1707–1783) was concerned about its true meaning in his book *Algebra* (1770), when he wrote: "Such numbers, which by their nature are impossible, are called *imaginary* or *fanciful numbers* [our italics] because they exist only in the imagination." It would take almost three centuries after they were first introduced (and used with great suspicion) before an adequate theory would be created that would embrace them in a wider "field of arithmetic" (Burton, 1991). The next section is devoted to such a field.

If you are willing to accept it on faith that no inconsistencies can arise in this very shallow treatment of complex numbers from high school, then you can skip the next section without missing a great deal. Just one thing though; consider the following example, which shows how the ordinary rules of algebra can run amuck. You might change your mind about skipping the next section. In this example, we assume that i exists, and that it equals $\sqrt{-1}$.

$$(-1)^2 = 1$$

$$(-1) \cdot (-1) = 1$$

$$\sqrt{(-1) \cdot (-1)} = \sqrt{1}$$

$$\sqrt{-1} \cdot \sqrt{-1} = 1$$

$$i \cdot i = 1$$

$$i^2 = 1$$

But this contradicts $i^2 = -1$! Was any law of algebra violated? It seems not. So what went wrong?

A FOUNDATION FOR THE COMPLEX NUMBERS

In order to better understand the discussion which follows, it is helpful to take a broader view of what constitutes a "number." This is no different from the task you once had when negative numbers were introduced in elementary algebra; at that point, you had worked exclusively with positive numbers and zero, and then suddenly you had to learn how to work with a new kind of number. But any system of objects that obeys certain operational working rules can be regarded as a system of "numbers." The operations of addition (+) and multiplication (·) are defined in some unambiguous manner which must obey certain standard rules. These rules are familiar from elementary algebra—we assume you know what they are formally. They include the commutative laws of + and · (such as $a \cdot b = b \cdot a$), the associative laws, and the distributive law of multiplication, among others. Such laws are the attributes required in mathematics in order to define what is generally known as a **field**. One can regard the members of any such field as *numbers* because they behave the same way as the real numbers do.

Thus, the real number system with all its familiar working rules is a field, denoted by $\mathbb{R}$. We aim to use this system to create a larger field which includes the number i, and more generally, the complex numbers $a + bi$. There are several ways to do this; we have chosen a system you should be familiar with, namely, 2×2 *matrices of real numbers*. In case this is new to you, let's describe what these objects are and how they work. This will at the same time provide review for others.

A real 2×2 *matrix* is merely a square array of numbers arranged in two rows and two columns. For example, the "numbers" in this system take on the appearance:

$$\begin{bmatrix} 1 & 2 \\ 5 & 4 \end{bmatrix}, \begin{bmatrix} 0 & -1 \\ 2 & 3 \end{bmatrix}, \begin{bmatrix} a & b \\ c & d \end{bmatrix}, \text{ and } \begin{bmatrix} x & y \\ z & w \end{bmatrix}$$

where $a, b, c, d, x, y, z,$ and w are real numbers. Next, as in a typical course in linear algebra, the *sum* and *product* of these matrices are defined:

$$\begin{bmatrix} a & b \\ c & d \end{bmatrix} + \begin{bmatrix} x & y \\ z & w \end{bmatrix} = \begin{bmatrix} a+x & b+y \\ c+z & d+w \end{bmatrix} \quad \text{and} \quad \begin{bmatrix} a & b \\ c & d \end{bmatrix} \cdot \begin{bmatrix} x & y \\ z & w \end{bmatrix} = \begin{bmatrix} ax+bz & ay+bw \\ cx+dz & cy+dw \end{bmatrix}$$

The product rule is more complicated than we might expect, but there are good reasons for it, having to do with topics in linear algebra (which will be omitted). The important thing is that since we can add and multiply such matrices, they begin to act like "numbers". And familiar rules emerge. For example, the associative laws of addition and multiplication are valid. The distributive law of multiplication is also valid: for any three matrices A, B, and C, $A(B + C) = AB + AC$. We are not going to take the trouble to prove all these laws since this is normally covered in linear algebra courses (the problem section at the end of this chapter will consider a few of these proofs). But it is important to realize that for matrices in general, the ordinary rules of arithmetic apply. One exception: the commutative law of multiplication is not always valid. Examples abound, like the following:

$$\begin{bmatrix} 1 & 2 \\ 3 & 4 \end{bmatrix}\begin{bmatrix} 1 & 1 \\ 1 & 1 \end{bmatrix} = \begin{bmatrix} 3 & 3 \\ 7 & 7 \end{bmatrix}, \qquad \begin{bmatrix} 1 & 1 \\ 1 & 1 \end{bmatrix}\begin{bmatrix} 1 & 2 \\ 3 & 4 \end{bmatrix} = \begin{bmatrix} 4 & 6 \\ 4 & 6 \end{bmatrix}$$

That is, AB is not always equal to BA, a significant departure from ordinary numbers. This property is required however, in order to obtain a field. What can be done about this? One way to solve this problem is to restrict the set of matrices we are willing to work with. For example, consider only matrices like

$$A = \begin{bmatrix} a & 0 \\ 0 & a \end{bmatrix}$$

If we multiply two such matrices in any order, we obtain

$$AB = \begin{bmatrix} a & 0 \\ 0 & a \end{bmatrix}\begin{bmatrix} b & 0 \\ 0 & b \end{bmatrix} = \begin{bmatrix} ab+0\cdot 0 & a\cdot 0+0\cdot b \\ 0\cdot b+a\cdot 0 & 0\cdot 0+a\cdot b \end{bmatrix} = \begin{bmatrix} ab & 0 \\ 0 & ab \end{bmatrix}$$

By interchanging a and b,

$$BA = \begin{bmatrix} b & 0 \\ 0 & b \end{bmatrix}\begin{bmatrix} a & 0 \\ 0 & a \end{bmatrix} = \begin{bmatrix} ba & 0 \\ 0 & ba \end{bmatrix} = AB$$

So the commutative law of multiplication is valid. What about the numbers "zero" and "one"? That's easy. Just define

$$O = \begin{bmatrix} 0 & 0 \\ 0 & 0 \end{bmatrix} \quad \text{and} \quad I = \begin{bmatrix} 1 & 0 \\ 0 & 1 \end{bmatrix}$$

Now all the operational rules that are valid for real numbers are valid for this class of matrices. For example, $A + O = A$, $OA = O$ and $IA = A$ for any matrix A. But we have not gotten very far, because if the ordinary rule for scalar multiplication be adopted as it is in linear algebra, then all our matrices are of the form aI, for real a, and this just becomes a fancy way to display the real numbers, where the matrix aI represents the real number a. Thus instead of obtaining $X^2 = -I$ for some matrix X, we find that the square of a matrix in this system is just

$$(aI)^2 = (aI)(aI) = \begin{bmatrix} a & 0 \\ 0 & a \end{bmatrix}\begin{bmatrix} a & 0 \\ 0 & a \end{bmatrix} = \begin{bmatrix} a^2+0\cdot 0 & a\cdot 0+0\cdot b \\ 0\cdot b+a\cdot 0 & 0\cdot 0+a^2 \end{bmatrix} = \begin{bmatrix} a^2 & 0 \\ 0 & a^2 \end{bmatrix} = a^2I$$

which represents the non-negative real number a^2. Thus, squaring a matrix of this form does not produce a negative number (or one that we would regard as a negative number).

It takes much experimentation to come up with a different system that includes a matrix which we can regard as i. It was discovered a long time ago that the desired system we are seeking consists of all matrices of the form

$$\begin{bmatrix} a & b \\ -b & a \end{bmatrix}$$

where a and b are any two real numbers. Not only does the commutative law of multiplication hold for these matrices (proof?), but other properties can be observed. We are going to let you experiment a little to see if some of the ordinary rules for a field are true.

Now suppose that $a = 0$ and $b = 1$. We shall denote the resulting matrix by J. That is,

$$J = \begin{bmatrix} 0 & 1 \\ -1 & 0 \end{bmatrix}$$

If we square this matrix, the result is

$$J^2 = \begin{bmatrix} 0 & 1 \\ -1 & 0 \end{bmatrix}\begin{bmatrix} 0 & 1 \\ -1 & 0 \end{bmatrix} = \begin{bmatrix} -1 & 0 \\ 0 & -1 \end{bmatrix} = (-1)I$$

Does this suggest anything? If $b = 0$ and a is any real number, we obtain the matrix

$$A = \begin{bmatrix} a & 0 \\ 0 & a \end{bmatrix} = aI$$

which we have identified with the real number a. Thus, our new system of matrices *includes all the real numbers*, and also a number J that acts like *i*. Indeed, if we define *i* as the matrix J, then, as above, $i^2 = J^2 = (-1)I = -1$.

You might want to explore this system further on your own, and to continue the development. It provides the very number system we were looking for, one that contains the real number system in a consistent manner, one that obeys all the desirable algebraic rules, and one that shows how *i* fits in algebraically. We do not have to invoke artificial rules or add axioms to produce desired algebraic properties. All the familiar rules of algebra (the *field properties*) hold for these 2×2 matrices. It constitutes the field of **complex numbers**, denoted by $\mathbb{C}$, a field that includes $\mathbb{R}$ as a subfield. (You are encouraged to work out the essential details of this system.)

EXAMPLE 1

Find the matrix products

$$\begin{bmatrix} 2 & 3 \\ -3 & 2 \end{bmatrix}\begin{bmatrix} 4 & 5 \\ -5 & 4 \end{bmatrix} \quad \text{and} \quad \begin{bmatrix} 4 & 5 \\ -5 & 4 \end{bmatrix}\begin{bmatrix} 2 & 3 \\ -3 & 2 \end{bmatrix}$$

and compare your answer with $(2 + 3i)(4 + 5i)$ as calculated in your high school algebra course.

SOLUTION

$$\begin{bmatrix} 2 & 3 \\ -3 & 2 \end{bmatrix}\begin{bmatrix} 4 & 5 \\ -5 & 4 \end{bmatrix} = \begin{bmatrix} 2\cdot4+3(-5) & 2\cdot5+3\cdot4 \\ -3\cdot4+2(-5) & -3\cdot5+2\cdot4 \end{bmatrix} = \begin{bmatrix} -7 & 22 \\ -22 & -7 \end{bmatrix}$$

and

$$\begin{bmatrix} 4 & 5 \\ -5 & 4 \end{bmatrix}\begin{bmatrix} 2 & 3 \\ -3 & 2 \end{bmatrix} = \begin{bmatrix} 4\cdot2+5(-3) & 4\cdot3+5\cdot2 \\ -5\cdot2+4(-3) & -5\cdot3+4\cdot2 \end{bmatrix} = \begin{bmatrix} -7 & 22 \\ -22 & -7 \end{bmatrix}$$

$$(2 + 3i)(4 + 5i) = 2\cdot4 + 2\cdot5i + 3i\cdot4 + 3i\cdot5i = 8 + 10i + 12i + 15i^2 = 8 + 22i - 15 = -7 + 22i.$$

GRAPHING COMPLEX NUMBERS—THE POLAR REPRESENTATION

An old name for a visual (geometric) treatment of complex numbers is the *Argand diagram*. Its construction was an attempt by mathematicians to explain complex numbers using geometry. The principal persons involved with this discovery (which was made independently at about the same time) were a Norwegian surveryor and cartographer Caspar Wessel (1745–1818), a French-Swiss bookkeeper Jean-Robert Argand (1768–1822), and Carl Friedrich Gauss (1777–1855); (Burton, 1991).

Since a complex number $z = x + iy$ consists of a unique pair of real numbers (x and y), it can be plotted as a point in the xy-plane. Thus, for example, $2 + i$, $3i$, and $-4 - 3i$ are represented by unique points, as shown in Figure 1.1. With this technique, a graphical representation of complex numbers emerges. The xy-plane may thus be referred to as the **complex plane**. The number $3i$, for example (called **pure imaginary**), lies on the positive y-axis 3 units from the origin. When z is real ($y = 0$), it lies on the x-axis (called the **real axis**), and when z is pure imaginary ($x = 0$), it lies on the y-axis (called the **imaginary axis**); all other complex numbers are located in one of the four quadrants.

THE COMPLEX PLANE

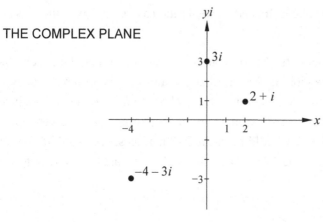

FIGURE 1.1

The arithmetic operations for complex numbers now take on a geometric meaning. Addition is actually vector addition, obeying the parallelogram law: If $z = x + iy$ and $w = u + iv$, then $z + w \equiv (x + u) + i(y + v)$. Thus the sum $z + w$ is the fourth vertex of the parallelogram having consecutive vertices z, O (the origin), and w, as illustrated in Figure 1.2, just as is the case for vector addition.

Another property of addition comes from this geometric interpretation. The **absolute value** of a complex number z, denoted by $|z|$, is defined to be the distance from the origin to z (sometimes called the *modulus* or *magnitude* of z). That is, by the Pythagorean theorem, $|z| = |x + iy| = \sqrt{x^2 + y^2}$. One can then observe from Figure 1.2 the important *triangle inequality* for complex nmbers

(1.1) $$|z + w| \leq |z| + |w|$$

In Figure 1.2, the length of the dashed line joining z and $z + w$ equals the distance from O to w, or $|w|$ (due to properties of parallelograms), so the geometric triangle inequality in the triangle having vertices O, z, and $z + w$ proves **(1.1)**.

Multiplication involves two geometric properties. First, the distance from O to the product zw is the product of the individual distances from O to z and from O to w. That is,

(1.2) $$|zw| = |z| \cdot |w|$$

Secondly, the angle which the ray from O to zw makes with the positive x-axis is the *sum* of the two angles θ and φ made by Oz and Ow individually, as in Figure 1.3. That is,

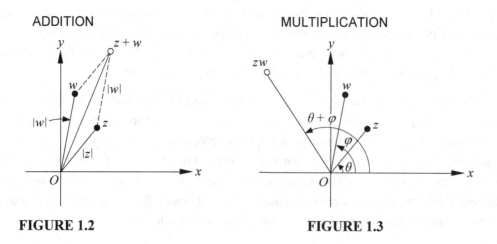

FIGURE 1.2   **FIGURE 1.3**

(1.3) $$\arg zw = \arg z + \arg w$$

where, in general, $\arg z$ denotes the **argument** of the complex number z, defined as the angle which the line Oz makes with the positive x-axis. This property is the first of many intriguing features of complex numbers to be revealed throughout this book.

Establishing **(1.2)** and **(1.3)** requires the use of *polar coordinates*. Recall this concept from algebra/trigonometry: each point $P(x, y)$ in the plane except the origin can be represented by polar coordinates (r, θ), where r is the signed distance from O to P, and θ is the measure of the angle from the positive x-axis to the ray OP (Figure 1.4). The number r can be negative if one identifies $(-r, \theta)$ with the point $(r, \theta + \pi)$, which lies on the ray opposite ray OP; and θ can be any multiple of 2π plus the initial angle

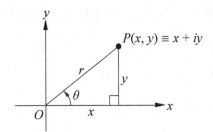

FIGURE 1.4

defined by ray OP. The following conversion formulas should be familiar to you, valid for all possible values of r and θ:

(1.4) $$x = r \cos \theta \qquad \text{and} \qquad y = r \sin \theta$$

Although r and θ can be any two real numbers, for definiteness we shall from now on restrict r to be positive and require θ to lie on the interval $(-\pi, \pi]$. If $(x, y) \neq (0, 0)$, the formulas **(1.4)** may then be inverted to produce:

(1.5) $$r = \sqrt{x^2 + y^2} \qquad \text{and} \qquad \theta = \tan^{-1} \frac{y}{x} + k\pi \quad (k = 0 \text{ or } \pm 1)$$

where the signs and value of k are chosen so that $-\pi < \theta \leq \pi$. For convenience, $\tan^{-1} y/x$ is taken as a symbol for (1) $\pi/2$ if $x = 0$ and $y > 0$, and (2) $-\pi/2$ if $x = 0$ and $y < 0$. Thus, in general, $\theta = \tan^{-1} y/x$ if $x \geq 0$, $\theta = \tan^{-1} y/x + \pi$ if $x < 0$ and $y \geq 0$, and $\theta = \tan^{-1} y/x - \pi$ if x and y are both negative (θ is not defined if x and y are both zero. With this restriction, we define θ as the **principle argument** of the complex number z, denoted $\text{Arg} z$.

Writing a complex number in **polar form** amounts to representing (x, y) in polar coordinates (r, θ), and substituting **(1.4)** into z. This produces the expression

$$x + iy = (r \cos \theta) + i(r \sin \theta) \equiv r \operatorname{cis} \theta$$

Note that the engineer's shorthand symbol $\operatorname{cis} \theta$ for $\cos \theta + i \sin \theta$ was used; this will occasionally appear in future work. The terms of the polar form of $x + iy$ may be written entirely in terms of x and y (in the case of the general polar form, the answers are not uniquely determined):

(1.6) General Polar Form for $z = x + iy$: $z = r \operatorname{cis} \theta$ where $r^2 = x^2 + y^2$ and $\tan \theta = y/x$.

(1.7) Restricted Polar Form: $z = r \operatorname{cis} \theta$ where $r = \sqrt{x^2 + y^2}$ and $\theta = \tan^{-1} \frac{y}{x} \pm k\pi$ (as in **(1.5)**)

EXAMPLE 2

Write the following complex numbers in restricted polar form, using special angles for (a) and (b), and using 5-place decimals for (c) and (d). (As in calculus, radians must be used for angle measure in all calculations.)

(a) $3 + 3i$

(b) $2\sqrt{3} + 2i$

(c) $3 - 5i$ (Quadrant II)

(d) $6.18934 + 0.95237i$ (Quadrant I)

SOLUTION

(a) Using a graph we can see that $\theta = 45°$ or $\pi/4$ radians, and that $r = \sqrt{18} = 3\sqrt{2}$. Hence

$$z = 3\sqrt{2}(\cos \pi/4 + i \sin \pi/4) = 3\sqrt{2} \operatorname{cis} \pi/4$$

(b) $r = \sqrt{12+4} = \sqrt{16} = 4$ and $\theta = \tan^{-1}2/2\sqrt{3} = \tan^{-1}\sqrt{3}/3 = 30°$ or $\pi/6$ radians, so that

$$z = 4(\cos \pi/6 + i \sin \pi/6) = 4 \operatorname{cis} \pi/6$$

(c) $r = \sqrt{9+25} = \sqrt{34}$ and $\theta = \tan^{-1}(-5/3) + \pi = 2.11122$. Hence,

$$z = \sqrt{34} \ (\cos 2.11122 + i \sin 2.11122) = 5.83095 \operatorname{cis} 2.11122$$

(d) $r = \sqrt{38.3079296+0.9070086} = \sqrt{39.2149382} = 6.26218;$ $\theta = \tan^{-1}0.95237/6.18934 = 0.15268$, which produces

$$z = 6.26218 \operatorname{cis} 0.15268$$

NUMERICAL EXPERIMENT

Consider the complex numbers $3 + 4i$ and $12 - 5i$. Write these numbers in polar form using 5-decimal accuracy for θ (and φ) as in Example 2(d). This leads to

$$3 + 4i = r \operatorname{cis} \theta \qquad \text{and} \qquad 12 - 5i = s \operatorname{cis} \varphi$$

(r and s are integers.) Next, take the product $(3 + 4i) \cdot (12 - 5i)$ using ordinary algebra, then convert the answer to polar form

$$(3 + 4i)(12 - 5i) = t \operatorname{cis} \omega$$

where ω is a 5-place decimal. Now test the product properties mentioned above to see if $t = rs$ and $\omega = \theta + \varphi$.

In terms of polar coordinates, the product of two complex numbers $z = r \operatorname{cis} \theta$ and $w = s \operatorname{cis} \varphi$ becomes

$$z \cdot w = (r \operatorname{cis} \theta)(s \operatorname{cis} \varphi) = rs \operatorname{cis} \theta \operatorname{cis} \varphi$$

In order to establish **(1.3)** we need to show that $\operatorname{cis} \theta \operatorname{cis} \varphi = \operatorname{cis}(\theta + \varphi)$. Using the algebra of complex numbers, note that by definition, $\operatorname{cis}\theta \cdot \operatorname{cis}\varphi$ is the same as

$$(\cos\theta + i\sin\theta)\cdot(\cos\varphi + i\sin\varphi) = \cos\theta\cos\varphi + i^2\sin\theta\sin\varphi +$$

$$i(\cos\theta\sin\varphi + \sin\theta\cos\varphi) = \cos\theta\cos\varphi - \sin\theta\sin\varphi + i(\cos\theta\sin\varphi + \sin\theta\cos\varphi)$$

Recall the addition identities for sine and cosine; one thus obtains

$$\operatorname{cis}\theta \cdot \operatorname{cis}\varphi = \cos(\theta + \varphi) + i\sin(\theta + \varphi) = \operatorname{cis}(\theta + \varphi)$$

We have thereby proved a very useful theorem in complex variables. It leads to a result known as *DeMoivre's theorem*, after Abraham de Moivre (1667–1754) who discovered it in 1730. (DeMoivre was better known for his extensive work in probability.) It plays such an important role in complex variables that a formal statement is warranted. First, we state formally the concepts established so far.

THEOREM 1.8

For any two complex numbers z and w, if $z = r\operatorname{cis}\theta$ and $w = s\operatorname{cis}\varphi$, then $zw = rs\operatorname{cis}(\theta + \varphi)$.

COROLLARY A

For any two complex numbers z and w, $|zw| = |z|\cdot|w|$.

Proof: Let $z = r\operatorname{cis}\theta$ and $w = s\operatorname{cis}\varphi$. By definition,

$$|z| = \sqrt{(r\cos\theta)^2 + (r\sin\theta)^2} = \sqrt{r^2(\sin^2\theta + \cos^2\theta)} = \sqrt{r^2} = r$$

since $r > 0$. Similarly, $|w| = s$. By Theorem 1.8, $|zw| = |rs\operatorname{cis}(\theta + \varphi)| = rs = |z\|w|.$ \\

COROLLARY B

For any two complex numbers z and w, $\arg zw = \arg z + \arg w$.

Proof: (See text following Numerical Experimental.) \\

Note that the above two corollaries establish the properties **(1.2)** and **(1.3)**. We finally come to DeMoivre's theorem.

COROLLARY C (DeMoivre's Theorem)

If θ is any real number and $p = m/n$ rational, $(\operatorname{cis}\theta)^p = \operatorname{cis}p\theta$.

[The relatively simple proof of this will be left for problems (see Problems 18–20). The first step is to show that $(\operatorname{cis}\theta)^2 = \operatorname{cis}2\theta$, derived from Theorem **1.8** (when $\theta = \varphi$), then to extend this to $(\operatorname{cis}\theta)^n = \operatorname{cis}n\theta$ for all integers n. Can you see how to do this using mathematical induction?]

This section will end with a list of some basic notions and standard notation used in complex variables: some of them have already appeared. Frequent use of these will be made as we proceed further in our study of complex variables. (Note that the argument of zero is undefined.)

(1.9)   *Real part* of a complex number: $\mathcal{R}(x + iy) = x$

(1.10)   *Imaginary part* of a complex number: $\mathcal{I}(x + iy) = y$

(1.11)   *Equality* of two complex numbers: $z = w$ iff $\mathcal{R}(z) = \mathcal{R}(w)$ and $\mathcal{I}(z) = \mathcal{I}(w)$

(1.12)   *Absolute value* (*modulus*) of a complex number: $|x + iy| = \sqrt{x^2 + y^2}$

(1.13)   *Argument* of a nonzero complex number: $\arg z \equiv \arg(x + iy) = \theta$, where (r, θ) are arbitrary polar coordinates for z (θ is any solution of $\tan\theta = y/x$). (NOTE: $\arg z$ is multivalued.)

(1.14)   *Principle argument* of a nonzero complex number: $\operatorname{Arg} z = \arg z$, where $\arg z$ is chosen on the interval $(-\pi, \pi]$. (NOTE: $\operatorname{Arg} z$ is uniquely determined as in **(1.5)**.)

(1.15)   *Triangle inequality*: For any two complex numbers z and w, $|z + w| \leq |z| + |w|$

(1.16)   *Polar form* of a complex number: $z = |z| \operatorname{cis}(\arg z) \equiv |z|(\cos\arg z + i \sin\arg z)$

(1.17)   *Restricted polar form* of a complex number: $z = |z| \operatorname{cis}(\operatorname{Arg} z) \equiv |z|(\cos\operatorname{Arg} z + i \sin\operatorname{Arg} z)$

(1.18)   *Complex conjugate* of a complex number: $\overline{z + w} = x - iy$

Properties: (1) $\overline{z + w} = \overline{z} + \overline{w}$
(2) $\overline{z \cdot w} = \overline{z} \cdot \overline{w}$
(3) $z \cdot \overline{z} = |z|^2$

PROBLEMS

1. Solve the equation $x^2 - 8x + 41 = 0$.

2. Check your answer in Problem 1 by direct substitution.

3. Consider the matrix

$$X = \begin{bmatrix} 4 & -5 \\ 5 & 4 \end{bmatrix}$$

Use the computation rules for matrices, show that X satisfies the matrix equation $X^2 - 8 \cdot X + 41 \cdot I = O$ by substitution. Convert this to complex numbers (where $X = 4 - 5i$), and compare.

4. The inverse of a 2×2 matrix A is any matrix B such that $AB = BA = I$ (which can be shown to be unique if it exists).

   (a) Show that if

$$A = \begin{bmatrix} a & b \\ c & d \end{bmatrix} \quad \text{and} \quad B = \frac{1}{k}\begin{bmatrix} d & -b \\ -c & a \end{bmatrix} \quad \text{where} \quad k = ad - bc$$

then B is the inverse of A, denoted A^{-1}.

(b) Show that if z is the matrix given by

$$z \equiv x + iy = \begin{bmatrix} x & y \\ -y & x \end{bmatrix}$$

then $z^{-1} \equiv 1/z$ exists.

5. Using the definition in Problem 4,

   (a) Use the matrix form for $3 - 4i$ to find $(3 - 4i)^{-1}$.

   (b) Using (a), calculate $(2 + 3i)/(3 - 4i)$ if this is defined to mean $(2 + 3i)[(3 - 4i)^{-1}]$.

   (c) Compare with the procedure you learned in algebra using complex conjugates:

   $$\frac{a + bi}{c + di} = \frac{a + bi}{c + di} \cdot \frac{c - di}{c - di} .$$

6. Prove the commutative law of multiplication in general for matrices of the type defined by $x + iy$:

   $$AC = \begin{bmatrix} a & b \\ -b & a \end{bmatrix} \begin{bmatrix} c & d \\ -d & c \end{bmatrix} = CA$$

7. Prove the distributive law for 2×2 matrices, $A(B + C) = AB + AC$, by computing the following:

   (a) $\begin{bmatrix} a & b \\ c & d \end{bmatrix} \begin{bmatrix} u+x & v+y \\ r+z & s+w \end{bmatrix}$ (representing $A(B + C)$)

   (b) $\begin{bmatrix} a & b \\ c & d \end{bmatrix} \begin{bmatrix} u & v \\ r & s \end{bmatrix}$ and $\begin{bmatrix} a & b \\ c & d \end{bmatrix} \begin{bmatrix} x & y \\ z & w \end{bmatrix}$ (representing AB and AC)

   (c) The sum of the two products you obtained in (b), representing $AB + AC$.

8. What matrix does $a + bi$ represent? Prove your result by using operations in matrices, starting with $aI + (bI)J$.

9. If you are familiar with the rules, show that $\mathcal{C}$ is a field.

10. Given $z = -1 + i$ and $w = 2 + 2i$, find

    (a) $|z|$ and $|w|$.

    (b) Polar form of z and w.

    (c) Examine the rule $\text{Arg} zw = \text{Arg} z + \text{Arg} w$.

    (d) Sketch the graph of z and w.

11. Prove the rules in (1.18) for complex conjugates (with $z = x + iy$ and $w = u + iv$).

12. Prove the triangle inequality without using geometry, where $z = a + bi$ and $w = c + di$. [**Hint:** This is equivalent to proving $(a + c)^2 + (b + d)^2 \le (a^2 + b^2) + 2\sqrt{(a^2 + b^2)(c^2 + d^2)} + (c^2 + d^2)$. Show details.]

13. Show that the product

    (*) $$(2 + 2i) \cdot \left(\frac{\sqrt{3} + 1}{2} + \frac{\sqrt{3} - 1}{2} i \right)$$

    is the same as

    $$(\sqrt{8} \text{ cis } 45°)(\sqrt{2} \text{ cis } 15°)$$

    Use ordinary algebra to find the product (*), and write it in polar form. Verify the rules (1) and (2) given above for products of complex numbers. Sketch the graph for this problem.

14. Repeat the analysis of Problem 13 for the product zw where $z = 6 + 2\sqrt{3}i$ and $w = -1 - \sqrt{3}i$.

15. Referring to Problem 13 for the polar form, show that $[(\sqrt{3} + 1)/2 + (\sqrt{3} - 1)i/2]^6 = 8i$ in two ways: By DeMoivre's theorem, then by direct multiplication (obtain c^2, then cube this result by the rule given in Table 1 to obtain $(c^2)^3 = c^6$).

16. Use DeMoivre's theorem to find $(1 + i)^6$.

17. Use DeMoivre's theorem to find $(1 - i)^{100}$.

18. Prove:

  (a) $(\operatorname{cis}\theta)^2 = \operatorname{cis}2\theta$ [That is, show that $(\cos\theta + i\sin\theta)^2 = \cos2\theta + i\sin2\theta$.]

  (b) Prove that if $(\operatorname{cis}\theta)^n = \operatorname{cis}n\theta$ for some positive integer n, then $(\operatorname{cis}\theta)^{n+1} = \operatorname{cis}(n+1)\theta$.

  (c) If you are familiar with mathematical induction, prove that (a) and (b) imply that $(\operatorname{cis}\theta)^n = \operatorname{cis}n\theta$ for all positive integers n.

19. Prove $(\operatorname{cis}\theta)^{-1} = \operatorname{cis}(-\theta)$, then show that $(\operatorname{cis}\theta)^n = \operatorname{cis}n\theta$ for all integers n.

20. Prove from Problem 18 that $(\operatorname{cis}\theta)^p = \operatorname{cis}p\theta$ for all rational numbers p ($= m/n$ for integers m and n). [**Hint:** First show that $(\operatorname{cis}\theta)^{1/n} = \operatorname{cis}(\theta/n)$ by setting $\varphi = \theta/n$ (thus $\theta = n\varphi$) and using Problem 18(c).]

21. The nth roots of unity. Consider the equation $z^n = 1$. This equation has n solutions. Find them by evaluating $\operatorname{cis}(2k\pi/n)$ for $k = 1, 2, \cdots, n$, and show that they satisfy the equation $z^n = 1$.

22. More generally, the equation $z^n = c$ can be written in polar form $z^n = r\operatorname{cis}\theta$ where $r > 0$. Take the nth root of both sides to one obtain

$$z = r^{1/n}\operatorname{cis}(\theta/n + 2k\pi/n), \quad \text{where } k = 1, 2, \cdots, n$$

which leads to the n solutions of $z^n = c$. Use this to find:

  (a) The three cube roots of i. Sketch the graph, locating these roots.

  (b) Find the four solutions of the equation $z^4 = 8 + 8\sqrt{3}i$. Sketch the graph.

23. As in Problem 22, find the fifth roots of $32i$ and plot these points in the complex plane. What do you observe?

24. A test for consistency. Solve the equation $z^3 = 1$ by factoring $z^3 - 1$ to find the three roots by ordinary methods in algebra, then show that this produces the same answers obtained by Problem 22 with $n = 3$.

25. Prove the following corollary to Corollary A of Theorem 1.2:

COROLLARY D

If $-\pi < \operatorname{Arg}z + \operatorname{Arg}w \leq \pi$, then $\operatorname{Arg}zw = \operatorname{Arg}z + \operatorname{Arg}w$.

[**Hint:** First show that if $\arg c \neq \operatorname{Arg}c$ then either $\arg c > \pi$ or $\arg c \leq -\pi$. Also, $\operatorname{Arg}z = \arg z$ by definition.]

26. Is the matrix version of $\mathbb{C}$ the only class of 2×2 matrices over the real numbers that produces a field?

27. Matrices and the Quaternions. A number system developed by W.R. Hamilton in 1843 is an example of an algebra which satisfies all the field properties except the commutative law of multiplication. A **quaternion** is a number of the form $a + bi + cj + dk$ where a, b, c, and d are reals and i, j, and k are pure imaginaries satisfying the rules $i^2 = j^2 = k^2 = -1$, $ij = -ji = k$, $jk = -kj = i$, and $ki = -ik = j$. The ordinary rules of algebra are taken axiomatically. [For example, $(3 + i)\cdot(1 - 2i + j) = 5 - 5i + 3j + k$.] Like complex numbers, the validity of the quaternions is in question. Consider $\mathbb{Q}$, the set of all 2×2 matrices over the complex numbers of the form $\begin{bmatrix} z & w \\ -\overline{w} & \overline{z} \end{bmatrix}$, where z and w are complex numbers. Next, define these matrices in $\mathbb{Q}$:

$$\mathrm{I}^* = \begin{bmatrix} 1 & 0 \\ 0 & 1 \end{bmatrix} \qquad \mathrm{I} = \begin{bmatrix} i & 0 \\ 0 & -i \end{bmatrix} \qquad \mathrm{J} = \begin{bmatrix} 0 & 1 \\ -1 & 0 \end{bmatrix} \qquad \mathrm{K} = \begin{bmatrix} 0 & i \\ i & 0 \end{bmatrix}$$

  (a) Show that the system $\mathbb{Q}$ is closed under matrix addition and matrix multiplication.

  (b) Show that $a\mathrm{I}^* + b\mathrm{I} + c\mathrm{J} + d\mathrm{K} = \begin{bmatrix} a+bi & c+di \\ -c+di & a-bi \end{bmatrix}$ and that $\mathrm{I}^2 = \mathrm{J}^2 = \mathrm{K}^2 = -\mathrm{I}^*$, $\mathrm{IJ} = -\mathrm{JI} = \mathrm{K}$, $\mathrm{JK} = -\mathrm{KJ} = \mathrm{I}$, and $\mathrm{KI} = -\mathrm{IK} = \mathrm{J}$. Thus, the matrix $a\mathrm{I}^* + b\mathrm{I} + c\mathrm{J} + d\mathrm{K}$ is an appropriate represention of $a + bi + cj + dk$.

  (c) Explore by numerical examples a few quaternion products and their matrix representations.

  (d) Show that $(a + bi + cj + dk)(a - bi - cj - dk) = a^2 + b^2 + c^2 + d^2$, and use this to find the multiplicative inverse of $a + bi + cj + dk$. For a matrix representation, use the formula for the inverse of a matrix.

THE ELEMENTARY FUNCTIONS

The elementary functions you studied in algebra and calculus can be extended to complex variables in such a way that many properties you are familiar with carry over to the complex field. These functions include the power function x^a for real a, the trigonometric functions ($\sin x$, $\cos x$, $\tan x$), the exponential and logarithmic functions (e^x, $\ln x$), and the hyperbolic functions ($\sinh x$, $\cosh x$, $\tanh x$). Our task is to construct these same functions for $\mathbb{C}$ in a logical manner. While this could be taken care of by simply presenting the definitions without explanation (an approach you may actually prefer), we are going to delve a little deeper into the *motivations* for making those definitions. This will provide a more interesting development. To keep us focused, our ultimate goal is to find a logical way to compute the complex number i^i. Does it even exist? One of the objects of this chapter will be to find an answer.

THE POWER FUNCTION

The easiest function to define is the power function $f(z) = z^p$, where p is a rational number. There is no problem when p is a positive integer n, because powers $z^n = (x + iy)^n$ follow from algebra and the binomial theorem (which is valid in any field, including $\mathbb{C}$). The first section in Chapter 1 indicated as much. But when p is not an integer, the definition for z^p may not be as clear.

Recall that for real variables, x^p is not defined for non-integral values of p unless $x \geq 0$. We might wonder if there is going to be a similar restriction in the case z^p for complex z. Note that in the special case $p = \frac{1}{2}$ for example, we are toying with the question of how the *square root* function $\sqrt{z} \equiv (x + iy)^{\frac{1}{2}}$ ought to be defined. As in algebra, it will be defined as some unique (complex) number w such that $w^2 = z$. It could be imagined that since we have taken care of the matter in the case of $\sqrt{-1}$ by the very act of constructing $\mathbb{C}$, it is possible that we might not need restrictions on z in defining $\sqrt{z}$, and more generally for z^p. We shall see.

Since the square root function stands out as an important special case, we devote a separate analysis for it. This will be done by solving the equation $w^2 \equiv (u + iv)^2 = z$ for w. Thus

$$(u + iv)^2 = z = x + iy \quad \Rightarrow \quad u^2 - v^2 + 2uv \cdot i = x + iy$$

Equating real and imaginary parts, we conclude that

$$u^2 - v^2 = x \quad \text{and} \quad 2uv = y$$

We must solve this system of equations for u and v in terms of x and y. Since $v = y/2u$ from the second equation (assume, momentarily, that $u \neq 0$), substitute this into the first equation to obtain $u^2 - (y/2u)^2 = x$. Thus we obtain the equation

$$u^2 - \frac{y^2}{4u^2} = x \qquad \text{or} \qquad 4u^4 - 4xu^2 - y^2 = 0$$

This is quadratic in u^2, so we can use the quadratic formula for real numbers to solve for u^2. (Note that the quadratic at right is valid when $u = 0$.) After simplifying,

$$u^2 = \frac{x + \sqrt{x^2 + y^2}}{2}$$

(Since $u^2 \geq 0$ and the radical expression is greater than x for $y > 0$, the plus sign in front of the radical must be chosen). Continue solving for u and v; we have (by rationalizing the denominator for v),

$$u = \pm\sqrt{\frac{\sqrt{x^2 + y^2} + x}{2}} \qquad \text{and} \qquad v = \frac{y}{2u} = \pm\sqrt{\frac{\sqrt{x^2 + y^2} - x}{2}}$$

Since the choice is ours to make, we *define* the **square root radical** of z by

(2.1)
$$\sqrt{z} \equiv \sqrt{x + iy} = \sqrt{\frac{\sqrt{x^2 + y^2} + x}{2}} + i\varepsilon\sqrt{\frac{\sqrt{x^2 + y^2} - x}{2}}$$

where $\varepsilon = 1$ if $y = 0$, and $\varepsilon = y/|y|$ otherwise. Note that if $z = x$ (real), then $\sqrt{z} = \sqrt{x}$ provided $x \geq 0$. Also, if $x = -1$ and $y = 0$, (2.1) produces the result $\sqrt{-1} = i$. More generally, $\sqrt{-a} = \sqrt{a}i$ for real $a > 0$.

   Observe the uniqueness of the answer, similar to $\sqrt{x}$ for $x \geq 0$ in $\mathbb{R}$. Note also that there are no restrictions on z. This enables us to solve any quadratic equation in $\mathbb{C}$ for precisely two solutions. (Both the method of completing-the-square and the quadratic formula are valid for $\mathbb{C}$ since the algebraic laws are the same for $\mathbb{R}$ and $\mathbb{C}$; thus the quadratic formula for complex numbers is the same as that for the reals).

EXAMPLE 1

Using (2.1), find
(a) $\sqrt{-9}$
(b) $\sqrt{3 - 4i}$
(c) $\sqrt{ai}$ where $a > 0$ (real)

SOLUTION
(a) $z = -9$ implies $x = -9$, $y = 0$; thus (2.1) becomes

$$\sqrt{-9} = \sqrt{\frac{\sqrt{9^2 + 0} + (-9)}{2}} + i\sqrt{\frac{\sqrt{9^2 + 0} - (-9)}{2}} = \sqrt{0} + i\sqrt{\frac{18}{2}} = 3i$$

(b) $x = 3$ and $y = -4$, which implies that $\sqrt{x^2 + y^2} = 5$ and $y/|y| = -1$, In this case,

$$\sqrt{3 - 4i} = \sqrt{\frac{5 + 3}{2}} - i\sqrt{\frac{5 - 3}{2}} = 2 - i$$

(c) Here, $x = 0$ and $y = a > 0$ so that

$$\sqrt{ai} = \sqrt{\frac{\sqrt{0+a^2}+0}{2}} + i\sqrt{\frac{\sqrt{0+a^2}-0}{2}} = \sqrt{\frac{a}{2}} + i\sqrt{\frac{a}{2}}$$

EXAMPLE 2

Solve the quadratic equation $z^2 + 2iz - (1 + 2i) = 0$.

SOLUTION

Using the quadratic formula,

$$z = \frac{-2i \pm \sqrt{(2i)^2 + 4(1+2i)}}{2} = \frac{-2i \pm \sqrt{-4+4+8i}}{2} = \frac{-2i \pm \sqrt{8i}}{2} = -i \pm (1+i)$$

[From Example 1(c), $\sqrt{8i} = 2 + 2i$.] Thus the two solutions are $z = 1$ and $z = -1 - 2i$.

For roots of higher order, and for all other rational exponents, a more general approach must be used. Consider the equation $w^n = z \equiv x + iy$, where n is an integer > 2. By De Moivre's theorem as extended to rational exponents (see Problem 20, Chapter 1), we can write, in polar form, $z^{1/n} = (r \operatorname{cis}\theta)^{1/n} = r^{1/n}\operatorname{cis}\theta/n$. That is (since we are concerned with the radical and not the general solution to $w^n = z$), define

(2.2) $\qquad \sqrt[n]{z} \equiv \sqrt[n]{x+iy} = (x^2 + y^2)^{1/2n}\operatorname{cis}(1/n)(\tan^{-1}y/x + k\pi) \qquad (z \neq 0)$

where k is defined as in **(1.5)**. More generally, for any rational number $n = p/q$, $q > 0$, define

(2.3) $\qquad \boxed{z^n = |z|^n\operatorname{cis}(n\operatorname{Arg}z) \qquad (z \neq 0)}$

This defines the **power function**, $f(z) = z^n$ for rational exponents n. For non-integral values of n ($= p/q$), the root **(2.3)** represents only one of the q solutions of the equation $w^q = z^p$ for w in terms of z. If n is a positive integer, **(2.3)** is in agreement with the algebraically-derived z^n for $n = 2, 3, 4, 5$, as presented at the beginning of Chapter 1. (Just for practice, you should try showing this explicitly for $n = 2$ and 3.) Thus there is only one choice for z^n when n is a positive integer; if n is a negative integer ($n = -m$ for $m > 0$), again $w = (z^{-m})^{1/n}$ is unique. Finally, if n is any rational number and $z = x$ is real, then $z^n = x^n$ for $x \geq 0$. (In this chapter, the elementary funcions will all have the desirable feature of reducing to their real counterparts when z is a real number.)

EXAMPLE 3

Use **(2.3)** to find $(3 + 4i)^{2/3}$ and express the answer in decimal form accurate to 5 places.

SOLUTION

$$(3+4i)^{2/3} = [5(\operatorname{cis}\tan^{-1}4/3)]^{2/3} = 25^{1/3}[\cos(2/3\tan^{-1}4/3) + i\sin(2/3\tan^{-1}4/3)] = 2.38285 + 1.69466i$$

Further discussion: It is interesting to make a numerical test of the above result. Let's cube this according to the rule in Table 1, Chapter 1, and see if this compares with $(3 + 4i)^2$. The real part of $(2.38285 + 1.69466i)^3$ is given by $(2.38285)^3 - 3(2.38285)(1.69466)^2 = 13.529761 - 20.52972 = -6.999959$, while the imaginary part is $3(2.38285)^2(1.69466) - (1.69466)^3 = 28.866707 - 4.86684 = 23.999860$, yielding $-7 + 24i$ (to 3-decimal accuracy), while $(3 + 4i)^2 = 9 - 16 + 2 \cdot 3 \cdot 4i = -7 + 24i$.

A few properties of the power function can be stated and proved. Proofs will be omitted here, but they are not too difficult. You should try your hand at them, making use of Theorem **1.8**). Restrictions are required in some cases. In particular, the rule $\sqrt{z}\sqrt{w} = \sqrt{zw}$ is not valid in general.

(2.4)   For rational m and n,
$$\begin{cases} z^0 = 1 \\ z^m z^n = z^{m+n} \\ z^m/z^n = z^{m-n} \\ (zw)^m = z^m w^m \\ (z^m)^n = z^{mn} \end{cases}$$

THE EXPONENTIAL FUNCTION

Do you remember the infinite series expansion for e^x? This will provide a reasonable method for defining e^{iy} for real y, which will lead to a good definition of e^z. It was the method Euler used in deriving his famous equation $e^{\pi i} + 1 = 0$ in 1740, which is said to connect the *five most important numbers in mathematics*. The infinite series goes like this:

$$e^x = 1 + x + \frac{x^2}{2!} + \frac{x^3}{3!} + \cdots + \frac{x^n}{n!} + \cdots$$

Suppose we just formally substitute iy for x in the above series without worrying too much about meaning or validity. Based on simple algebra and using the rules established for i, the result is

$$e^{iy} = 1 + iy + \frac{i^2 y^2}{2!} + \frac{i^3 y^3}{3!} + \cdots + \frac{i^n y^n}{n!} + \cdots$$
$$= \left(1 - \frac{y^2}{2!} + \frac{y^4}{4!} + - \cdots \pm \frac{y^p}{p!} + \cdots\right) + i\left(y - \frac{y^3}{3!} + \frac{y^5}{5!} - + \cdots \pm \frac{y^q}{q!} + \cdots\right) \quad (p \text{ even}, q \text{ odd})$$

Note that the infinite series inside the first set of parentheses is the series for $\cos y$, while that in the second set is the series for $\sin y$. Thus, we *define*, along with Euler,

(2.5) $$\boxed{e^{iy} = \cos y + i \sin y \equiv \text{cis}\, y}$$

[Euler's equation is a direct result of **(2.5)**.]

The above presentation does not constitute a proof since it involves infinite sums of complex variables not yet justified. The above can only be regarded as *motivation* for making the definition **(2.5)**. It constitutes the style of Euler, who often did not worry about details of proof. However, one can actually make this development rigorous. So the definitions we make here are not entirely arbitrary. In fact it can be shown that they produce *the only functions possible for complex z that coincide with their real counterparts when z is real* (assuming also that the functions are differentiable).

A direct consequence of **(2.5)** is an alternate way to write a complex number in polar form:

(2.6) $$z = r(\cos\theta + i \sin\theta) = re^{i\theta}$$

Now that we have defined e^{iy}, the definition for $e^z \equiv e^{x+iy}$ suggests itself. A formal application of a law of exponents yields

$$e^{x+iy} = e^x e^{iy} = e^x(\cos y + i \sin y)$$

Thus, it makes sense to define

(2.7)

$$e^z \equiv e^{x+iy} = e^x(\cos y + i \sin y) = e^x \text{cis} y$$

This definition is valid for all complex numbers z. It defines the **exponential function** for complex variables, often denoted $\exp(z)$. Note the property $e^z = e^x$ for real z (when $y = 0$).

EXAMPLE 4

Use **(2.7)** to find, in standard $a + bi$ form, expressing your answer in decimal form:

(a) e^{3+2i}

(b) e^{1-i}

(c) If $z = 3 + 2i$ and $w = 1 - i$, show by direct calculation that $e^z \cdot e^w = e^{z+w}$.

SOLUTION

(a) $e^{3+2i} = e^3(\cos 2 + i \sin 2) \approx e^3(-0.4161468 + 0.9092974i) \approx -8.358533 + 18.263727i$

(b) $e^{1-i} = e(\cos 1 - i \sin 1) \approx e(0.5403023 - 0.8414710i) \approx 1.468694 - 2.287355i$

(c) $e^z \cdot e^w = (-8.358532 + 18.263727i) \cdot (1.468694 - 2.287355i) = (-12.276126 + 41.775627) + (19.118930 + 26.823826)i = 29.49950 + 45.94276i$, while $e^{z+w} = e^{4+i} = e^4(\cos 1 + i \sin 1) = e^4(0.5403023 + 0.8414710i) = 29.49951 + 45.94276i$, and we find agreement.

Conclusive evidence that we have made a good definition for the exponential function emerges from the following mathematical result.

THEOREM 2.8

Let z and w be any two complex numbers, and let n be any rational real number. Then the following is true:

(a) $e^0 = 1$

(b) $e^z e^w = e^{z+w}$

(c) $e^z/e^w = e^{z-w}$

(d) $(e^z)^n = e^{nz}$ (n real, rational)

Proof: (The proofs of (a) and (c) will be left for problems.) For (b), let $z = x + iy$ and $w = u + iv$. Then $e^z e^w \equiv e^{x+iy} e^{u+iv} = (e^x \text{cis} y) \cdot (e^u \text{cis} v) = e^{x+u} \text{cis} y \cdot \text{cis} v = e^{x+u} \text{cis}(y + v) = e^{x+u+i(y+v)} = e^{z+w}$. To prove (d), we use De Moivre's theorem as extended to rational powers: $(e^z)^n = (e^x \text{cis} y)^n = (e^x)^n (\text{cis} y)^n = e^{nx} \text{cis} ny = e^{nx+iny} = e^{n(x+iy)} = e^{nz}$. $\backslash\backslash$

A final result is found to be useful later. It is a direct result of the definition for e^z. It simply states that if $-\pi < y \le \pi$, the absolute value of the complex number e^{x+iy} is e^x, and its principle argument is y. (You should ask what happens if either $y > \pi$ or $y \le -\pi$ and try to give an answer.)

THEOREM 2.9

For all complex numbers $z = x + iy$ such that $-\pi < y \le \pi$, $|e^z| = e^x$ and $\text{Arg } e^z = y$.

Proof: By definition,

$$|e^z| = |e^x(\cos y + i \sin y)| = [(e^x \cos y)^2 + (e^x \sin y)^2]^{1/2} = [e^{2x}(\cos^2 y + \sin^2 y)]^{1/2} = e^x$$

In restricted polar form, $e^z = |e^z|(\cos \theta + i \sin \theta) = e^x(\cos \theta + i \sin \theta)$ where $\theta = \text{Arg} e^z$.

Thus $e^x \mathrm{cis}\, y = e^x \mathrm{cis}\, \theta$ or $\mathrm{cis}\, y = \mathrm{cis}\, \theta$. Since both y and θ are in the range $(-\pi, \pi]$, it follows that $y = \theta = \mathrm{Arg}\, e^z$. ◊

THE TRIGONOMETRIC FUNCTIONS

The *circular functions* sine and cosine were no doubt introduced to you geometrically in terms of the unit circle in the xy-plane. In complex numbers, a non-geometric approach is necessary. First, by considering a formal application of the addition identity for the sine function, one observes

$$(*) \qquad \sin z \equiv \sin(x + iy) = \sin x \cos iy + \cos x \sin iy$$

This suggests that we first find a way to define $\cos iy$ and $\sin iy$. We do this by working with the definition already made for e^{iy}. Since $e^{iy} = \cos y + i \sin y$ for all real y, note the following analysis (where we make the substitution of $-y$ for y in the second step):

$$e^{iy} = \cos y + i \sin y$$

$$e^{-iy} = \cos y - i \sin y$$

$$e^{iy} + e^{-iy} = 2 \cos y$$

Now, substitute iy for y to obtain, again formally,

$$e^{i(iy)} + e^{-i(iy)} = 2 \cos iy \qquad \text{or} \qquad e^{-y} + e^{y} = 2 \cos iy$$

Recalling from calculus the definition for $\cosh y$, we are led to define

$$\cos iy = \tfrac{1}{2}(e^{y} + e^{-y}) = \cosh y$$

for all real y. In a similar manner, it follows that we should define

$$\sin iy = i \sinh y$$

(this will be left as a problem, namely, Problem 13 below.) Making substitutions into $(*)$, we obtain

$$\sin(x + iy) = \sin x \cos iy + \cos x \sin iy = \sin x \cosh y + \cos x (i \sinh y)$$

Thus we define the **sine function** as

(2.10)
$$\boxed{\sin z \equiv \sin(x + iy) = \sin x \cosh y + i \cos x \sinh y}$$

To define the cosine function first write (by analogy) $\cos(x + iy) = \cos x \cos iy - \sin x \sin iy$; thus, since $\cos iy = \cosh y$ and $\sin iy = i \sinh y$, we define **cosine** as

(2.11)
$$\boxed{\cos z \equiv \cos(x + iy) = \cos x \cosh y - i \sin x \sinh y}$$

For $\tan z$, as in real variables, let

(2.12)
$$\boxed{\tan z = \frac{\sin z}{\cos z} \qquad (\cos z \neq 0)}$$

These definitions all reduce to their real counterparts when z is real, and they make good sense for arbitrary z, as shown above. But they may seem rather strange, and this can lead to some unusual results. One of them is the pair of identities (which are traditionally taken as definitions):

(2.13)
$$\sin z = \frac{e^{iz} - e^{-iz}}{2i} \qquad \cos z = \frac{e^{iz} + e^{-iz}}{2}$$

You will be given the chance to prove this in Problem 14, an interesting challenge for you at this point. More familiar identities are, for complex numbers z and w:

$$\sin^2 z + \cos^2 z = 1, \qquad \cos(z + w) = \cos z \cos w - \sin z \sin w, \qquad \cos 2z = 2\cos^2 z - 1$$

EXAMPLE 5

Evaluate each of the following, and write the answer in decimal form, accurate to 5 places:
(a) $\sin i$
(b) $\cos(\pi + 3i)$

SOLUTION
(a) $\sin i = \sin 0 \cosh 1 + i \cos 0 \sinh 1 = 0 + i \sinh 1 = 1.17520i$
(b) $\cos(\pi + 3i) = \cos \pi \cosh 3 - i \sin \pi \sinh 3 = -\cosh 3 = -10.06766$.

The hyperbolic functions are somewhat easier to define—we can just copy the real variable definitions (since e^z has meaning for complex z):

(2.14)
$$\sinh z = \frac{e^z - e^{-z}}{2} \qquad \text{and} \qquad \cosh z = \frac{e^z + e^{-z}}{2}$$

with $\tanh z = \sinh z / \cosh z$ for all z.

THE LOGARITHM FUNCTION

The last major function to be considered is the logarithm. A logical definition emerges from **(2.6)**. Since the polar form of z is $re^{i\theta}$, we are led to hypothesize

$$\ln re^{i\theta} = \ln r + \ln e^{i\theta} = \ln r + i\theta \ln(e) = \ln r + i\theta$$

where ordinary properties of logarithms are applied formally. Since θ is not uniquely defined for arbitrary complex numbers, we use the *principle argument of z*. Thus the **logarithm function** may be defined as

(2.15)
$$\ln z = \ln |z| + i \, \text{Arg} \, z$$

for z any complex number $\neq 0$. This function can be expressed explicitly in terms of x and y by using a previous formulation for Arg z from Chapter 1:

(2.16) $\ln z \equiv \ln (x + iy) = \ln \sqrt{x^2 + y^2} + i(\tan^{-1}y/x + k\pi)$ $(z \neq 0)$

as in **(1.5)**. The definition is, explicitly,

(2.16′) $\ln z \equiv \ln(x + iy) = \begin{cases} \ln \sqrt{x^2 + y^2} + i\tan^{-1}\dfrac{y}{x} & \text{if } x > 0 \\[2ex] \ln \sqrt{x^2 + y^2} + i \cdot \dfrac{y}{|y|} \cdot \dfrac{\pi}{2} & \text{if } x = 0 \\[2ex] \ln \sqrt{x^2 + y^2} + i(\tan^{-1}\dfrac{y}{x} + \pi) & \text{if } x < 0, y \geq 0 \\[2ex] \ln \sqrt{x^2 + y^2} + i(\tan^{-1}\dfrac{y}{x} - \pi) & \text{if } x < 0, y < 0 \end{cases}$

Unlike the real-valued logarithm function which is not defined for $x < 0$, observe that if $z = x < 0$, then $\ln z = \ln|x| + \pi i$. In fact, in complex variables, $\ln z$ *is defined for all complex numbers* $z \neq 0$.

EXAMPLE 6

Calculate

(a) $\ln i$ exactly

(b) $\ln c$ where $c = (-3 - 2i)$ and write the answer in decimal form to 5-place accuracy

(c) $e^{\ln c}$ using the result in (b). Show that $e^{\ln c} = c$ to within 5-place accuracy.

SOLUTION

(a) By definition, $\ln i = \ln\sqrt{1} + i(\pi/2) = \pi i/2$.

(b) $\ln(-3 - 2i) = \ln \sqrt{13} + i (\tan^{-1}2/3 - \pi) = 1.28247 - 2.55359i$.

(c) $e^{\ln c} = e^{1.28247-2.55359i} = e^{1.28247}[\text{cis}(-2.55359)] = 3.605534[\cos(-2.55359) + i\sin(-2.55359)] = -2.99999 - 1.99999i$ (approximately $-3 - 2i$).

The major properties of logarithms for real numbers are (1) $\ln xy = \ln x + \ln y$ for $x, y > 0$, and (2) $\ln x^y = y \ln x$ for $x > 0$. Under certain restrictions, property (1) follows for the complex case, but the proof of (2) must wait until we define the general power function for complex variables (next section).

THEOREM 2.17

If $\text{Arg}\,z + \text{Arg}\,w$ lies on the interval $(-\pi, \pi]$, then $\ln zw = \ln z + \ln w$.

Proof: By Corollary D in Chapter 1 (Problem 25),

$$\text{Arg}\,zw = \text{Arg}\,z + \text{Arg}\,w$$

Therefore,

$$\ln zw = \ln|zw| + i\,\text{Arg}\,zw$$
$$= \ln|z||w| + i(\text{Arg}\,z + \text{Arg}\,w)$$
$$= (\ln|z| + i\,\text{Arg}\,z) + (\ln|w| + i\,\text{Arg}\,w)$$
$$= \ln z + \ln w. \;\;\diagdown$$

Recall that in real variables the functions $\ln x$ and $\exp x$ are *inverse functions*. Thus, $\ln(\exp x) = x$ for all real x, and $\exp(\ln x) = x$ for all $x > 0$. (That is, $\ln e^x = x$ and $e^{\ln x} = x$.) The same result is valid for the complex variable functions $\ln z$ and $\exp z$, provided z is restricted appropriately. We must obtain the domain and range of the logarithm function in order to address this issue satisfactorily.

NOTE: The general concept for inverse functions in mathematics is the following: if f maps D onto R one-to-one in the xy-plane, then f^{-1} exists, mapping R to D, where $f^{-1}(f(x)) = x$ for all $x \in D$, and $f(f^{-1}(y)) = y$ for all $y \in R$. ◈

The domain D of the logarithm function is {all complex $z \neq 0$}—the so-called *punctured* or *deleted* *plane*. Its range R interpreted in the xy-plane is the parallel strip $-\infty < x < \infty$, $-\pi < y \leq \pi$, as seen from **(2.15)**. On D one can observe that $\ln z$ is one-to-one. Thus, we know that $\ln^{-1} z$ exists, which may or may not equal e^z at this point. On the other hand, the domain of e^z is all complex z, and its range is all $z \neq 0$ (since $\operatorname{cis} y \neq 0$ for real y). But e^z is not one-to-one on its domain. As in the case of real variables (notably for the trigonometric functions), it is often necessary to restrict the domain of a function in order to define its inverse. Likewise, here we restrict the domain of e^z to the same parallel strip R as above ($-\infty < x < \infty$, $-\pi < y \leq \pi$), and e^z is found to be one-to-one on R. (See Figure 2.1 for illustration.) Thus, we see that $\ln z$ maps D to R one-to-one, and e^z maps R to D one-to-one. The fact that e^z and $\ln z$ are inverse functions for complex variables is established in the theorem which follows.

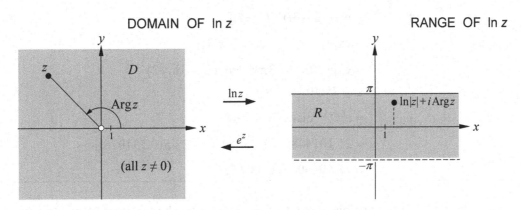

FIGURE 2.1

THEOREM 2.18

For all $z \in D$, $e^{\ln z} = z$, and for all $z \in R$, $\ln e^z = z$.

Proof: Letting $z \in D$ have the polar form $r \operatorname{cis} \theta$, where $r > 0$ and $\theta = \operatorname{Arg} z$,

$$e^{\ln z} = e^{\ln r + i\theta} = e^{\ln r}\operatorname{cis} \theta = r \operatorname{cis} \theta = z \,.$$

On the other hand, suppose that $z \equiv x + iy \in R$. The condition $-\pi < y \leq \pi$ is then satisfied; hence by Theorem **2.9**, $|e^z| = e^x$ and $\operatorname{Arg} e^z = y$. Thus,

$$\ln e^z = \ln|e^z| + i\operatorname{Arg} e^z = \ln e^x + iy = x + iy = z \ ◈$$

ARBITRARY POWERS OF COMPLEX NUMBERS

The exponential and logarithm functions can now be used to define z^w for arbitrary complex numbers z and w. You might recall from algebra or calculus that $a^b = \exp(\ln a^b) = \exp(b \ln a)$. This enables one to calculate mathematically such powers as $3^{\sqrt{2}}$ and π^e, for example. Although not as important as previous topics, it is desirable to develop this concept for complex variables. This will make it possible to calculate $z^w \equiv (a+bi)^{c+di}$ for virtually all complex numbers $z = a + bi$ and $w = c + di$.

With $\exp z$ defined as e^z, since all the necessary definitions have been made, we state:

(2.19) Power of a Complex Number $\boxed{z^w = \exp(w \ln z) \qquad (z \neq 0)}$

EXAMPLE 7

Calculate i^i.

SOLUTION

By **(2.19)**, $i^i = \exp(i \ln i) = \exp[i(\ln|i| + i \operatorname{Arg} i)] = \exp[i(0 + i \cdot \pi/2) = \exp(-\pi/2) = e^{-\pi/2}$ which is a real number!

EXAMPLE 8

Calculate $(1 + i)^{2-3i}$, expressing the answer in $a + bi$ decimal form, accurate to 5 places.

SOLUTION

$$(1 + i)^{2-3i} = \exp[(2 - 3i)\ln(1 + i)]$$

$$= \exp[(2 - 3i)(\ln\sqrt{2} + i \cdot \pi/4)]$$

$$= \exp[2\ln\sqrt{2} + 3\pi/4 + i(\pi/2 - 3\ln\sqrt{2})]$$

$$= e^{\ln 2 + 3\pi/4 + i(\pi/2 - \frac{1}{2}\ln 8)}$$

$$= e^{\ln 2 + 3\pi/4}\operatorname{cis}(\pi/2 - \tfrac{1}{2}\ln 8)$$

$$= 21.101448[\cos(0.531076) + i\sin(0.531076)]$$

$$= 18.19499 + 10.68707i$$

Further comment: In the above example, it is interesting to test the above power definition against the usual exponent law $a^n \cdot a^m = a^{n+m}$. By observing the product $(1 + i)^2(1 + i)^{-3i}$ (which should equal $(1 + i)^{2-3i}$, we first obtain $(1 + i)^2 = 1^2 - 1^2 + 2i = 2i$. By **(2.19)**, $(1 + i)^{-3i} = \exp[-3i\ln(1 + i)] = \exp[-3i(\ln\sqrt{2} + i \cdot \pi/4)] = \exp(3\pi/4 - i \cdot \frac{1}{2}\ln 8) = e^{3\pi/4}\operatorname{cis}(-\frac{1}{2}\ln 8) = 5.343531 - 9.097497i$. Thus, the desired product is $2i \cdot (5.343531 - 9.097497i) = 18.19499 + 10.68706i$, in agreement with the result in Example 8.

OPTIONAL ANALYSIS: LAWS OF EXPONENTS

In order to obtain the laws of exponents for complex numbers analogous to those for real variables (e.g., $a^{b+c} = a^b \cdot a^c$ for $a \geq 0$ and b and c real numbers), one must restrict $w \ln z$ to the range R of the logarithm function $\ln z$; that is, we require that $-\pi < \mathcal{I}(w \ln z) \leq \pi$, as above. If this is done, then, among other desirable properties, $z^{w_1} \neq z^{w_2}$ if $w_1 \neq w_2$, for all z such that $\mathcal{I}(z) \equiv y \in (-\pi, \pi]$. We can thus prove:

THEOREM 2.20

If $z \neq 0$, then

(a) $z^{w_1} = z^{w_2}$ iff $w_1 = w_2$, provided $\mathcal{I}(w \ln z) \in (-\pi, \pi]$ for $w = w_k$, $k = 1, 2$, and $z \neq 1$,

(b) $z^{w_1} z^{w_2} = z^{w_1 + w_2}$.

Proof:

(a) If $z^{w_1} = z^{w_2}$ then $\exp(w_1 \ln z) = \exp(w_2 \ln z)$; since $w_1 \ln z$ and $w_2 \ln z$ lie in the domain for which $\exp w$ is one-to-one, it follows that $w_1 \ln z = w_2 \ln z$. Since $z \neq 1$, $w_1 = w_2$.

(b) By definition and (b) of Theorem **2.8**,

$$z^{w_1} z^{w_2} = \exp(w_1 \ln z) \cdot \exp(w_2 \ln z) = \exp(w_1 \ln z + w_2 \ln z) = \exp[(w_1 + w_2) \ln z] = z^{w_1 + w_2}. \;\backslash\!\backslash$$

THEOREM 2.21

If $z \neq 0$ and $\Im(w \ln z) \in (-\pi, \pi]$ for $w = w_1$ and $w = w_2$, then:

(a) $(z^{w_1})^{w_2} = (z^{w_2})^{w_1} = z^{w_1 w_2}$

(b) $\ln z^w = w \ln$

Proof:

(a) $(z^{w_1})^{w_2} = \exp(w_2 \ln z^{w_1}) = \exp[w_2 \ln (\exp (w_1 \ln z))] = \exp[w_2(w_1 \ln z)] = z^{w_1 w_2}$.

(b) Since $\ln z$ and $\exp z$ are inverse functions, $\ln z^w = \ln(\exp w \ln z) = w \ln z. \;\backslash\!\backslash$

In the problem section which follows, a counterexample for the general application of property (a) above is explored. The key point in the proof of (a) was $\ln \exp(w_1 \ln z) = w_1 \ln z$ which comes from the inverse property $\ln e^w = w$. This requires $w \ln z \in R$ (range of $\ln z$) and it is the same condition mentioned previously, namely $-\pi < \Im(w \ln z) \leq \pi$.

The precise region for which $-\pi < \Im(w \ln z) \leq \pi$ is complicated and difficult to determine. A simple requirement is for both z and w to lie in the annular ring $\frac{3}{8} \leq |z| \leq \frac{3}{4}$, as shown in Figure 2.2 (a), but other conditions are sufficient as well. Two examples are the following conditions (and the regions described by them):

(b) $\qquad\qquad\qquad \frac{3}{4} \leq |z| \leq e \quad$ and $\quad |w| \leq \frac{3}{4} \qquad$ [Figure 2.2(b)]

(c) $\qquad\qquad\qquad \frac{1}{2} \leq |z| \leq 2, \; 0 \leq \operatorname{Arg} z \leq \pi/4 \quad$ and $\quad w \in S \qquad$ [Figure 2.2(c)]

where S is the square bounded by the coordinate axes and the lines $x = 2$ and $y = -2$.

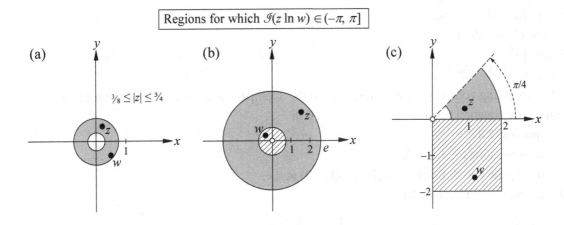

Figure 2.2

EXAMPLE 9

Show that $-\pi < \Im(w \ln z) \le \pi$ holds if both z and w lie in the annular ring $\frac{3}{8} \le |z| \le \frac{3}{4}$.

SOLUTION

We need to obtain the product $w \ln z$ explicitly. Let $z = x + iy$ and $w = u + iv$; then

$$w \ln z = (u + iv)(\ln|z| + i\operatorname{Arg} z) = u \ln|z| - v\operatorname{Arg} z + i(u\operatorname{Arg} z + v \ln|z|)$$

The imaginary part is thus given by $u\operatorname{Arg} z + v \ln|z|$, and the desired condition becomes

$$-\pi < u\operatorname{Arg} z + v \ln|z| \le \pi$$

By hypothesis, $\frac{3}{8} \le |z| \le \frac{3}{4}$, $|\operatorname{Arg} z| \le \pi$, $|u| \le \frac{3}{4}$, and $|v| \le \frac{3}{4}$. Hence $\ln\frac{3}{8} \le \ln|z| \le \ln\frac{3}{4} < 0$ and $|\ln|z|| \le |\ln\frac{3}{8}|$. By the triangle inequality,

$$|u\operatorname{Arg} z + v \ln|z|| \le |u\operatorname{Arg} z| + |v\ln|z|| = |u||\operatorname{Arg} z| + |v||\ln|z||$$

$$\le \tfrac{3}{4}\pi + \tfrac{3}{4}|\ln\tfrac{3}{8}| < 2.356 + 0.736 = 3.092 < \pi$$

which implies

$$-\pi < u\operatorname{Arg} z + v \ln|z| \le \pi$$

as desired.

PROBLEMS

1. Using **(2.1)**, show that if $z = x = \pm a$ (real), $a > 0$, the square root of z is
   (a) $\sqrt{a}$ if $x = a$,
   (b) $i\sqrt{a}$ if $x = -a$.
2. Calculate $\sqrt{5 + 12i}$ and $\sqrt[4]{5 + 12i}$.
3. Calculate $\sqrt{15 - 8i}$ and $\sqrt[4]{15 - 8i}$.
4. Using **(2.2)** and **(2.3)**, calculate:
   (a) $i^{-1/3}$, $i^{1/3}$, and $i^{2/3}$.
   (b) Verify that $i^{2/3} \cdot i^{-1/3} = i^{2/3 - 1/3}$.
   (c) $\sqrt[4]{15 - 8i}$ expressing your answer in 5-place decimals. Compare with your answer in Problem 3.
5. Solve the quadratic equation $z^2 - (3 + 2i)z + (5 + i) = 0$ in .
6. Prove the properties **(2.4)**.
7. Establish Euler's equation $e^{\pi i} + 1 = 0$.
8. Show that if $n = \frac{1}{2}$ the two definitions in **(2.1)** and **(2.3)** agree.
9. Prove parts (a) and (c) of Theorem **2.8**.
10. Calculate in $\mathbb{C}$:
    (a) $\cos \pi/2$
    (b) $\cos(\pi/2 + 2i)$
    (c) $\sin 2i$
11. Show that $\sin(\pi/2 + i\ln 3) = 5/3$.

12. Calculate

    (a) $\sinh 2i$

    (b) $\sinh(2+3i)$ (express the answer in $a+bi$ decimal form, accurate to 5 decimals).

13. Using the same method we used above for showing $\cos iy = \cosh y$, obtain the result $\sin iy = i\sinh y$.

14. Prove the identities **(2.13)**.

15. Prove the identity $\cos(z+w) = \cos z \cos w - \sin z \sin w$ in complex variables. [**Hint:** Start with the right side of the equation using the definitions for $\cos z$, $\cos w$, $\sin z$, and $\sin w$.]

16. Show that if $\sinh z = 0$ then $z = k\pi i$ where k is an integer. [**Hint:** From **(2.14)**, $e^{2z} = 1$.]

17. Show that although $\cosh x \neq 0$ for real x, there exist complex numbers z such that $\cosh z = 0$.

18. Show by direct calculation that

    (a) $\ln e^{4+3i} = 4 + 3i$

    (b) $\ln e^{3+4i} \neq 3 + 4i$

    (c) Resolve the inconsistency in (a) and (b).

19. Calculate 3^i.

20. Use **(2.19)** to find i^{i+1} and verify the relation $i^{i+1} = i^i \cdot i$. Note: It was established in Example 7 that $i^i = e^{-\pi/2}$.

21. Show that $(i^2)^i = (i^i)^2$ by direct calculation using **(2.19)**. Note: From Example 7, $i^i = e^{-\pi/2}$.

22. **(a)** Show that (c) of Theorem **2.21** fails by letting $z = i$, $w_1 = 1 + i$, and $w_2 = 3 + 3i$.

    (b) Show that in this case, $w_2 \ln z$ does not lie in the appropriate region.

23. Prove that if z and w lie in the second and third regions illustrated in Figure 2.2 (b) and (c), then

    $-\pi < w \ln z \leq \pi$.

24. Show that if $-\pi < \text{Arg } z + \text{Arg } w \leq \pi$, then for $q \neq 0$,

    (2.22) $\qquad\qquad\qquad\qquad (zw)^{1/q} = z^{1/q} w^{1/q} \qquad\qquad$ (q real)

    (Thus if q is an integer ≥ 2, the usual rule for radicals is valid under this restriction: $\sqrt[q]{zw} = \sqrt[q]{z}\sqrt[q]{w}$. For $q = 2$, discuss the counterexamples $z = w = -1$ and $z = w = -1 + i$ if this rule were applied in general.

25. **Differentiation and the Power Function.** Recall that $\partial f/\partial x$ is the *partial derivative* of the two-variable function $f(x, y)$ with respect to x (holding y constant). For example, $\partial/\partial x[x^4 y^2 + 5y] = 4x^3 y^2$. In complex variables, if $f(x, y) = u(x, y) + iv(x, y)$, we define $\partial f/\partial x$ as $\partial u/\partial x + i\partial v/\partial x$. In the expansion for z^4, the real and imaginary parts are, respectively, $u = x^4 - 6x^2 y^2 + y^4$ and $v = 4x^3 y - 4xy^3$.

    (a) Show that if $f(x, y) = z^4 \equiv u + iv$, then $\partial/\partial x \, (z^4) = 4z^3$.

    (b) Try out the idea in (a) when $f(x, y)$ equals z^2 and z^3 to see what you get.

26. As in Problem 24, experiment with $\partial f/\partial y$ where $f(x, y) = z^4$. You will discover that this does not agree with $4z^3$. However, a better result will occur if you compute $-i\partial f/\partial y$ instead. Thus, in this case, $\partial f/\partial x = -i\partial f/\partial y$. Try this out for $f(x, y) = z^3$ and $f(x, y) = z^2$.

27. In general, with $f(x, y) = u(x, y) + iv(x, y)$, find

    (a) $\partial f/\partial x$

    (b) $-i \, \partial f/\partial y$.

    (c) Use **(1.11)** to see what is required in order for the two complex partial derivatives of $f(z)$ to be equal. [If equality occurs, one is inclined to define $f'(x, y)$ ("total derivative") to be the common answer.]

    NOTE: The equations you should obtain in (c) are known as the *Cauchy-Riemann* conditions in complex variables, to be formally introduced and discussed in the next chapter. ◊

28. Referring to the Note above, show that the Cauchy-Riemann conditions are satisfied for

    (a) $f(z) = e^z$

    (b) $f(z) = \ln z$ for $\Im z > 0$ [if $\Re z < 0$ (use **2.16**)].

    (c) Find $\partial f/\partial x$ in each case, and simplify. Do these derivatives agree with those for real variables?

3

THE DERIVATIVE IN
COMPLEX VARIABLES

You may recall that in calculus the derivative was defined as a limit. Thus limits must be addressed before the derivative makes sense. The same is true for complex variables. We introduce this concept for complex variables before defining the derivative. Since this will essentially duplicate concepts you had in calculus, some of the details will be omitted. Following limits and continuity, the derivative concept for complex numbers is introduced, and this becomes the focus of attention. Many formulas are identical to those you had in calculus, but a few extra ideas will emerge as we will discover.

LIMITS FOR COMPLEX NUMBERS

The complex number z is said to *approach c*, or to come *arbitrarily close to c*, iff given real $\delta > 0$, $|z - c| < \delta$ for certain z. This is written symbolically $z \rightarrow c$. Note that $|z - c|$ is the distance from z to c in the complex plane, so we are requiring that, geometrically, z is close to c in terms of distance. As in real numbers, a function $f(z)$ is said to *approach L as z approaches c* provided that $|f(z) - L|$ becomes arbitrar-

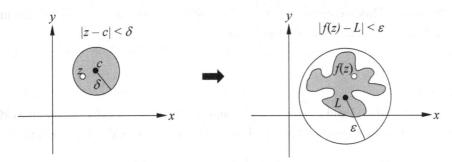

FIGURE 3.1

ily small as $|z - c|$ becomes arbitrarily small. Unlike limits in real variables which are restricted to one dimension, the values of z and $f(z)$ range throughout the xy-plane in some way, making it necessary to include a separate diagram, as in Figure 3.1. In complex variables, one cannot graph a function in the usual sense. Thus, the formal definition for limits in complex variables has 2-dimensional implications:

> **DEFINITION**   The limit of $f(z)$ is the complex number L as z approaches c if and only if (iff) given real $\varepsilon > 0$, there exists a real $\delta > 0$ such that if $0 < |z - c| < \delta$), then $|f(z) - L| < \varepsilon$).

NOTATION: If the limit of $f(z)$ exists and equals L as z approaches c, we write $\lim_{z \to c} f(z) = L$, as in real variables. Alternatively, the limit relation will often be written: $f(z) \to L$ as $z \to c$.

For example, if $f(z) = (7z - 6)/(5z + 15)$, then as z approaches -2, $7z - 6$ approaches -20, while $5z + 15$ approaches 5, and $f(z)$ evidently approaches $-20/5 = -4$. But, as in limits in real variables, there is much more to the concept of limits than this simplistic example. Limits often involve the *ratio* of two complex functions $f(z)/g(z)$, where both $f(z)$ and $g(z)$ approach zero and as $z \to c$. This leads to the *indeterminant form* 0/0, which, of course, cannot be evaluated directly. An example is the limit

$$\lim_{z \to 8} \frac{\sqrt[3]{z} - 2}{z - 8}$$

Here, an algebraic simplification seems impossible, and the only way to get some idea of this limit is to actually substitute values of z close to 8. If we do this, we obtain a table of values like the following [with $f(z) = (\sqrt[3]{z} - 2)/(z - 8)$]:

z	$f(z)$
7.5	0.085132
8.5	0.081655
7.9	0.083683
8.1	0.829885
7.99	0.083368
8.01	0.083299
7.999	0.083337
8.001	0.083330

(Note that we should also use values like $8.001 + 0.002i$ in order to properly examine this limit in the complex plane.)

It appears that the values of $f(z)$ are getting close to $0.0833333\cdots$ or $1/12$. Indeed, this is correct (the details are similar to those for the real variable limit corresponding to this problem).

Since limits for complex variables can be broken down and reduced to limits of real-valued functions, the basic theory is the same. This is embodied in the following result, which allows us to use our knowledge of limits for real variable functions to find limits in $\mathbb{C}$.

THEOREM 3.1

Suppose that $f(z) = u(x, y) + iv(x, y)$ and $c = a + bi$. The limit of $f(z)$ as $z \to c$ exists iff the individual limits of $u(x, y)$ and $v(x, y)$ exist as $x \to a$ and $y \to b$. That is, if $u(x, y) \to U$ and $v(x, y) \to V$, then

$$(3.1) \qquad \lim_{z \to c} f(z) = \lim_{x + iy \to a + bi} [u(x, y) + iv(x, y)] = \lim_{\substack{x \to a \\ y \to b}} u(x, y) + i \lim_{\substack{x \to a \\ y \to b}} v(x, y) = U + iV$$

Proof: First observe from the geometric definition in terms of distances, $z = x + iy \to a + ib = c$ iff $x \to a$ and $y \to b$. Hence, suppose that $u(x, y) \to U$ and $v(x, y) \to V$ as $x \to a$ and $y \to b$, and that $L = U + iV$. Then

$$|f(z) - L| = |u(x, y) + iv(x, y) - (U + iV)| = \sqrt{(u(x, y) - U)^2 + (v(x, y) - V)^2} \to 0$$

proving that $f(z) \to U + iV$. Conversely, suppose $f(z) \to L$ as $z \to c$ for some complex number $L =$

$U + iV$. Then

$$|u(x, y) + iv(x, y) - (U + iV)| = \sqrt{(u(x, y) - U)^2 + (v(x, y) - V)^2}$$

which converges to zero as $x \to a$ and $y \to b$. Thus $u(x, y) \to U$ and $v(x, y) \to V$. ⧅

EXAMPLE 1

Suppose that $f(z) = x \sin y + iy^2 \cos x$. Then since $u = x \sin y$ and $v = y^2 \cos x$ are continuous 2-variable functions of x and y,

$$\lim_{\substack{x \to \pi/2 \\ y \to \pi}} f(z) = \lim_{\substack{x \to \pi/2 \\ y \to \pi}} (x \sin y + iy^2 \cos x) = \pi/2 \sin \pi + i\pi^2 \cos \pi/2 = 0$$

EXAMPLE 2

Find the limit

$$\lim_{z \to 2+3i} \left[\frac{3xy - 9x}{y^2 - 9} + i\frac{2x^2 + x - 10}{x - 2} \right]$$

SOLUTION

Factoring enables us to eliminate the 0/0 form in each case:

$$\lim_{x \to 2, y \to 3} \left[\frac{3xy - 9x}{y^2 - 9} + i\frac{2x^2 + x - 10}{x - 2} \right] = \lim_{x \to 2, y \to 3} \left[\frac{3x(y - 3)}{(y + 3)(y - 3)} + i\frac{(2x + 5)(x - 2)}{x - 2} \right]$$

$$= \lim_{x \to 2, \, y \to 3} \left[\frac{3x}{y + 3} + i(2x + 5) \right]$$

$$= 1 + 9i$$

EXAMPLE 3

Evaluate the following limit:

$$\lim_{z \to 3+2i} \frac{z^3 - 2iz^2 - (9 + 6i)z}{z - (3 + 2i)}$$

SOLUTION

Factoring out the "z" in the numerator leads to a quadratic, which, if the limit exists, should contain the factor $z - (3 + 2i)$. To find the other factor, let's divide:

$$
\begin{array}{r}
z + 3 \\
z - (3 + 2i) \overline{\smash{\big)}\, z^2 - 2iz - (9 + 6i)} \\
\underline{z^2 - 3z - 2iz } \\
3z - (9 + 6i) \\
\underline{3z - (9 + 6i)} \\
\end{array}
$$

Thus,

$$\lim_{z \to 3+2i} \frac{z^3 - 2iz^2 - (9 + 6i)z}{z - (3 + 2i)} = \lim_{z \to 3+2i} \frac{z(z + 3)[z - (3 + 2i)]}{z - (3 + 2i)}$$

$$= \lim_{z \to 3+2i} z(z + 3)$$

$$= (3 + 2i)(6 + 2i) = 14 + 18i$$

The theorems on limits are proved exactly as they are for real numbers, so we state them without proof (actually, one can use Theorem **3.1** and the limit theorems from real analysis for easy proofs).

CONSTANT MULTIPLE THEOREM
If $\lim f(z) = L$ as $z \to c$ and a is any complex number, then $\lim af(z) = aL$.

SUM THEOREM
If $\lim f(z) = L$ and $\lim g(z) = M$ as $z \to c$, then $\lim [f(z) + g(z)] = L + M$.

PRODUCT THEOREM
If $\lim f(z) = L$ and $\lim g(z) = M$ as $z \to c$, then $\lim [f(z)g(z)] = LM$.

QUOTIENT THEOREM
If $\lim f(z) = L$ and $\lim g(z) = M \neq 0$ as $z \to c$, then $\lim [f(z)/g(z)] = L/M$.

CONTINUOUS FUNCTIONS

Continuity of functions in $\mathbb{C}$ closely resembles the real-variable concept. The formal definition is as follows (like its real counterpart):

DEFINITION   A function f in complex variables is said to be **continuous at** z_0 iff z_0 is in the domain of f and if for each real $\varepsilon > 0$ there exists $\delta > 0$ such that if $|z - z_0| < \delta$, then $|f(z) - f(z_0)| < \varepsilon$. A function f is **continuous in a region** R iff f is continuous at each point z_0 in R.

If one compares this definition with the one for limits, it is quickly concluded that $f(z)$ is continuous at z_0 iff $f(z_0)$ exists and

$$\lim_{z \to z_0} f(z) = f(z_0)$$

Thus the next theorem follows routinely.

THEOREM 3.2
If $f(z) \equiv f(x + iy) = u(x, y) + iv(x, y)$, then f is continuous at $z_0 = x_0 + iy_0$ iff each of the functions $u(x, y)$ and $v(x, y)$ are continuous at the point (x_0, y_0).

Important properties of continuous functions are that the sums, products, and quotients of two or more continuous functions are continuous. That is

$$f(z) + g(z), \quad f(z) \cdot g(z), \quad \text{and} \quad \frac{f(z)}{g(z)} \quad [\text{where } g(z) \neq 0]$$

are continuous functions provided $f(z)$ and $g(z)$ are continuous. These properties all follow directly from the corresponding sum, product, and quotient theorems on limits. A straightforward conclusion is that any polynomial $a_0 + a_1z + a_2z^2 + \cdots + a_nz^n$ is continuous, and any rational function (the quotient of two polynomials) is continuous at any point for which the denominator $\neq 0$. That is

$$\frac{a_0 + a_1z + a_2z^2 + \cdots + a_nz^n}{b_0 + b_1z + b_2z^2 + \cdots + b_mz^m}$$

is continuous for all z such that $b_0 + b_1 z + b_2 z^2 + \cdots + b_m z^m \neq 0$.

EXAMPLE 4

For what values z is the rational function $\dfrac{z^2 - 2iz + 1 - 4i}{z^3 + z^2 + 4z + 4}$ continuous?

SOLUTION

The quotient theorem on limits applies for all points except the zeroes of the denominator. To find them, we use group factoring:

$$z^3 + z^2 + 4z + 4 = z^2(z + 1) + 4(z + 1) = (z + 1)(z^2 + 4) = 0$$

with roots $z = -1$ and $z = \pm 2i$. Hence, the given function is continuous for all $z \in \mathbb{C}$ such that $z \neq -1, \pm 2i$.

Since these ideas are all rather straightforward, there is nothing new to learn here, except to adjust to the new generality required in dealing with complex numbers. In this connection, there is a very significant matter to point out, however. Although the idea of a complex number z that varies and approaches c as limit seems simple enough analytically, one must remember that instead of the familiar one-dimensional real number line, complex variable limits involve the *complex plane*—a two-dimensional environment. There can be many paths along which z (or a sequence of z's) can approach c geometrically, as illustrated in Figure 3.2.

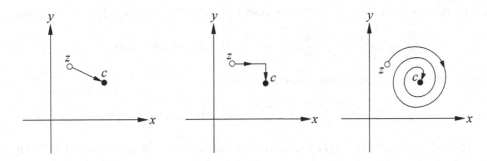

FIGURE 3.2

COMPLEX DIFFERENTIATION

The derivative for complex functions is the first instance where calculus in $\mathbb{C}$ begins to differ significantly from the calculus you learned in $\mathbb{R}$. Although we use the same basic definition, the consequences are more involved than one might expect, mainly due to the 2-dimensional property of complex numbers.

DEFINITION   A function f is said to be **differentiable** at z iff the limit $\lim\limits_{h \to 0} \dfrac{f(z + h) - f(z)}{h}$ exists, which is denoted $f'(z)$, or sometimes as $\dfrac{df}{dz}$. If $f'(z)$ exists for all z in a region R, the function $f'(z)$ is called the **derivative** of $f(z)$ on R.

EXAMPLE 5

Find the derivatives of **(a)** $f(z) = z^3$ and **(b)** $g(z) = z^{-1}$.

SOLUTION

(a)
$$f'(z) = \lim_{h \to 0} \frac{f(z+h) - f(z)}{h} \equiv \lim_{h \to 0} \frac{(z+h)^3 - z^3}{h}$$

$$= \lim_{h \to 0} \frac{z^3 + 3z^2h + 3zh^2 + h^3 - z^3}{h}$$

$$= \lim_{h \to 0} (3z^2 + 3zh + h^2) = 3z^2$$

(b)
$$g'(z) = \lim_{h \to 0} \frac{g(z+h) - g(z)}{h} \equiv \lim_{h \to 0} \frac{1}{h}\left(\frac{1}{z+h} - \frac{1}{z}\right)$$

$$= \lim_{h \to 0} \frac{1}{h}\left(\frac{z - (z+h)}{z(z+h)}\right) = \lim_{h \to 0} \frac{1}{h}\left(\frac{-h}{z(z+h)}\right)$$

$$= \lim_{h \to 0} \frac{-1}{z(z+h)} = -z^{-2}$$

Although the results in Example 5 seem to indicate that there are no significant modifications needed for differentiation in complex variables, the next result shows an aspect that is totally new. It is one of the most important concepts in complex variables. By breaking the function f into its real and imaginary parts, one can predict entirely in terms of those parts alone whether f is differentiable. Since it requires a knowledge of partial derivatives from calculus, it might be a good idea to review that concept.

Recall that the partial derivative of $u(x, y)$ with respect to x is the real variable limit, if it exists,

$$\frac{\partial u}{\partial x} = \lim_{h \to 0} \frac{u(x+h, y) - u(x, y)}{h} \qquad \text{(holding } y \text{ constant)}$$

Similarly, the partial derivative of $u(x, y)$ with respect to y is

$$\frac{\partial u}{\partial y} = \lim_{h \to 0} \frac{u(x, y+h) - u(x, y)}{h} \qquad \text{(holding } x \text{ constant)}$$

These concepts are needed in order to analyze the derivative concept introduced above. If we split $f(z)$ into its real and imaginary parts, the definition for $f'(z)$ becomes (with $z = x + iy$ and $h = h_1 + ih_2$)

$$\lim_{h \to 0} \frac{f(z+h) - f(z)}{h} = \lim_{h \to 0} \frac{f[x+h_1 + i(y+h_2)] - f(x+iy)}{h_1 + ih_2}$$

$$= \lim_{h \to 0} \frac{[u(x+h_1, y+h_2) + iv(x+h_1, y+h_2)] - [u(x,y) + iv(x,y)]}{h_1 + ih_2}$$

In the above limit the variable h is not restricted in any way, except that it must approach zero. It can start anywhere, and vary on any path that leads to zero. Also, since the limit is required to exist, it must be the same for all possible paths for h. Suppose we first consider the path lying on the x-axis ($h_2 = 0$). Then

$$\lim_{h \to 0} \frac{f(z+h) - f(z)}{h} = \lim_{h_1 \to 0} \frac{[u(x+h_1, y) + iv(x+h_1, y)] - [u(x,y) + iv(x,y)]}{h_1}$$

$$= \lim_{h_1 \to 0} \frac{u(x+h_1, y) - u(x, y)}{h_1} + i \lim_{h_1 \to 0} \frac{v(x+h_1, y) - v(x, y)}{h_1}$$

$$= \frac{\partial u}{\partial x} + i\frac{\partial v}{\partial x}$$

If the same procedure is used as h approaches 0 along the y-axis ($h = 0 + h_2 i$), we obtain

$$\lim_{h \to 0} \frac{f(z+h)-f(z)}{h} = \lim_{h_2 \to 0} \frac{[u(x,y+h_2)+iv(x,y+h_2)]-[u(x,y)+iv(x,y)]}{ih_2}$$

$$= \lim_{h_2 \to 0} \frac{u(x,y+h_2)-u(x,y)}{ih_2} + i \lim_{h_2 \to 0} \frac{v(x,y+h_2)-v(x,y)}{ih_2}$$

$$= \frac{1}{i}\frac{\partial u}{\partial y} + \frac{\partial v}{\partial y}$$

$$= \frac{\partial v}{\partial y} - i\frac{\partial u}{\partial y}$$

Comparing this result with that of the above, one obtains

$$\frac{\partial u}{\partial x} + i\frac{\partial v}{\partial x} = \frac{\partial v}{\partial y} - i\frac{\partial u}{\partial y}$$

Since the real and imaginary parts must agree, we conclude:

THEOREM 3.3 (Differentiation and the Cauchy-Riemann Conditions)
If the complex function $f(z) \equiv f(x + iy) = u(x, y) + iv(x, y)$ is differentiable, then the partial derivatives of u and v exist and must satisfy the conditions

(3.3) $$\frac{\partial u}{\partial x} = \frac{\partial v}{\partial y} \quad \text{and} \quad \frac{\partial u}{\partial y} = -\frac{\partial v}{\partial x}$$

NOTE: These conditions appeared as early as 1825 and are called the **Cauchy-Riemann** (CR) conditions, named after their discoverers, A. Cauchy and B. Riemann. ⧄

COROLLARY
If $f(z)$ is differentiable, then its derivative is given by

(3.4) $$f'(z) = \frac{\partial u}{\partial x} + i\frac{\partial v}{\partial x}$$

Proof: The limit defining the derivative was shown above to satisfy the relation

$$f'(z) = \lim_{h \to 0} \frac{f(z+h)-f(z)}{h} = \frac{\partial u}{\partial x} + i\frac{\partial v}{\partial x} \quad ⧄$$

EXAMPLE 6
The function $f(z) = z^2 \equiv x^2 - y^2 + i \cdot 2xy$ so that $u = x^2 - y^2$ and $v = 2xy$. Thus it can be seen that the CR conditions are satisfied (can you verify this?). Then evidently $f'(z)$ exists. Is this valid logically?

Example 6 involves the *converse* of Theorem 3.3. If one assumes that the partial derivatives involved are continuous, then the converse does hold. The next discussion is devoted to proving this converse result. It is optional material for you, but the final result is very important since it is one of the key concepts of complex variables. It will be shown that if $f(z) \equiv u(x, y) + iv(x, y)$ and the partial derivatives of $u(x, y)$

and $v(x, y)$ are continuous and satisfy the CR conditions, then $f(z)$ is differentiable.

The proof makes use of an *approximation theorem* in complex variables, analogous to the real-variable approximation theorem for functions of two variables. From multivariable calculus we know that a given two-variable function $u(x, y)$ having continuous partial derivatives $\partial u / \partial x$ and $\partial v / \partial y$ can be approximated by the formula

(3.5) $$u(x + h, y + k) = u(x, y) + \frac{\partial u}{\partial x} h + \frac{\partial u}{\partial y} k + \varepsilon_1 h + \varepsilon_2 k \qquad (\varepsilon_1, \varepsilon_2 \to 0 \text{ as } h, k \to 0)$$

For example, suppose $u(x, y) = x^2 y$. Then, using only basic algebra,

$$u(x + h, y + k) = (x + h)^2 (y + k) = (x^2 + 2xh + h^2)(y + k)$$

$$= x^2 y + x^2 k + 2xhy + 2xhk + h^2 y + h^2 k$$

$$= u(x, y) + (2xy)h + (x^2)k + (hy)h + (2xh + h^2)k$$

$$= u(x, y) + \frac{\partial u}{\partial x} h + \frac{\partial u}{\partial y} k + \varepsilon_1 h + \varepsilon_2 k$$

where $\varepsilon_1 = hy$ and $\varepsilon_2 = 2xh + h^2$. Thus, as $h, k \to 0$, then $\varepsilon_1, \varepsilon_2 \to 0$. For sake of completeness (and you may not have seen this argument), a proof of (3.5) is taken up in Problem 13 below, which shows why the continuity of the partial derivatives is needed.

It would be interesting to establish a similar approximation theorem for $\mathbb{C}$. Indeed, if the CR conditions hold and the partial derivatives for u and v are continuous, such a theorem can be obtained.

THEOREM 3.6 (Approximation Theorem for Complex Variables)
Let $f(z) = u(x, y) + iv(x, y)$, and suppose that the CR conditions hold for u and v and that the partial derivatives of u and v are continuous. Then for each $z = x + iy$ and complex number h

(3.6) $$f(z + h) = f(z) + \left(\frac{\partial u}{\partial x} + i \frac{\partial v}{\partial y} \right) h + \delta h$$

where δ is a function of h whose limit is zero as $h \to 0$.

In order to prove this, let's determine $f(z + h)$ in terms of $u(x, y)$, $v(x, y)$, and $h = h_1 + ih_2$. Assume that $f(z) = u(x, y) + iv(x, y)$ for all complex numbers $z = x + iy$. Then

$$f(z + h) = f(x + iy + h_1 + ih_2) = f(x + h_1 + i(y + h_2)) = u(x + h_1, y + h_2) + iv(x + h_1, y + h_2)$$

Using the approximation theorem (3.5) for both $u(x, y)$ and $v(x, y)$, there exist real numbers ε_1, ε_2, ε_1', and ε_2' with zero limits as $h_1, h_2 \to 0$ such that

$$f(z + h) = u(x + h_1, y + h_2) + iv(x + h_1, y + h_2) =$$

$$u(x, y) + \frac{\partial u}{\partial x} h_1 + \frac{\partial u}{\partial y} h_2 + \varepsilon_1 h_1 + \varepsilon_2 h_2 + i\left(v(x, y) + \frac{\partial v}{\partial x} h_1 + \frac{\partial v}{\partial y} h_2 + \varepsilon_1' h_1 + \varepsilon_2' h_2 \right)$$

$$= f(z) + \frac{\partial u}{\partial x} h_1 + \frac{\partial v}{\partial y} (ih_2) + i\left(\frac{\partial v}{\partial x} h_1 - \frac{\partial u}{\partial y} (ih_2) \right) + \varepsilon_1 h_1 + \varepsilon_2 h_2 + i(\varepsilon_1' h_1 + \varepsilon_2' h_2)$$

Now if we assume the CR conditions, we obtain (with $\delta_1 = \varepsilon_1 + i\varepsilon_1'$ and $\delta_2 = \varepsilon_2 + i\varepsilon_2'$)

$$f(z + h) - f(z) = \frac{\partial u}{\partial x} h_1 + \frac{\partial u}{\partial x} (ih_2) + i\left(\frac{\partial v}{\partial x} h_1 + \frac{\partial v}{\partial x} (ih_2) \right) + \delta_1 h_1 + \delta_2 h_2$$

$$= \left(\frac{\partial u}{\partial x} + i \frac{\partial v}{\partial x} \right) h_1 + \left(\frac{\partial u}{\partial x} + i \frac{\partial v}{\partial x} \right)(ih_2) + \delta_1 h_1 + \delta_2 h_2$$

$$= \left(\frac{\partial u}{\partial x} + i \frac{\partial v}{\partial x} \right) (h_1 + ih_2) + \delta h$$

where $\delta h = \delta_1 h_1 + \delta_2 h_2$. In Problem 15, you will be asked to show that $\delta \rightarrow 0$ as $h \rightarrow 0$. This proves **(3.6)**.

The approximation theorem places us only one step away from the derivative of $f(z)$. As before, let $f(z) = f(x + iy) = u(x, y) + iv(x, y)$, and suppose that the partial derivatives of u and v exist and are continuous, and that the CR conditions hold. By **(3.6)** we have

$$f(z + h) - f(z) = \left(\frac{\partial u}{\partial x} + i \frac{\partial v}{\partial x} \right) h + \delta h$$

Simply divide both sides by h and take the limit as $h \rightarrow 0$. Then $\delta \rightarrow 0$ and $f'(z)$ exists and equals $\partial u/\partial x + i \partial v/\partial x$. This proves the following result.

THEOREM 3.7 (Differentiation of Complex Functions)

If $f(z) \equiv f(x + iy) = u(x, y) + iv(x, y)$ and the functions u and v have continuous partial derivatives, then f is differentiable iff the CR conditions hold. Moreover, the following formula for the derivative is valid:

$$\textbf{(3.7)} \qquad f'(z) \equiv \frac{d}{dz} f(z) = \frac{\partial u}{\partial x} + i \frac{\partial v}{\partial x} = \frac{\partial u}{\partial y} - i \frac{\partial v}{\partial y}$$

Before considering examples, we define an important term in complex variables. In real analysis a function having a series expansion of f at a point x_0 is termed *analytic*, and this implies that all derivatives of f exist not only at x_0, but at every point in some open interval (a, b) about x_0 (called a *neighborhood* of x_0). The term *analytic* for complex variables is defined as follows.

DEFINITION   A function $f(z)$ in complex variables is said to be **analytic at** z_0 iff its derivative exists for all z in a neighborhood *of* z_0, that is, $f'(z)$ exists for all z such that $|z - z_0| < r$, for some positive r.

NOTE: A *neighborhood* of z_0 in the complex plane can be interpreted geometrically as the interior of a circle with center z_0 and nonzero radius r. Another frequently-used term for *analytic* is **holomorphic**. ⟍

EXAMPLE 7

Show that the CR conditions hold for the function $\ln z$ for all z not lying on the negative real axis and $z \neq 0$, and find its derivative.

SOLUTION

From **(2.16)**, $\ln z \equiv \ln (x + iy) = \ln \sqrt{x^2 + y^2} + i(\tan^{-1}y/x + k\pi)$ where k is constant at points above, or below, the negative x-axis [see **(2.16)**]. Here, $u = \ln \sqrt{x^2 + y^2}$ and $v = \tan^{-1}y/x + k\pi$. Thus,

$$\frac{\partial u}{\partial x} = \frac{\partial}{\partial x}[½\ln(x^2 + y^2)] = ½ \frac{1}{x^2 + y^2}(2x) = \frac{x}{x^2 + y^2} \quad \text{and} \quad \frac{\partial u}{\partial y} = \frac{\partial}{\partial y}[½\ln(x^2 + y^2)] = \frac{y}{x^2 + y^2}$$

Moreover,

$$\frac{\partial v}{\partial x} = \frac{\partial}{\partial x}(\tan^{-1}\frac{y}{x} + k\pi) = \frac{1}{1 + (y/x)^2} \cdot \frac{-y}{x^2} = \frac{-y}{x^2 + y^2} = -\frac{\partial u}{\partial y}$$

and

$$\frac{\partial v}{\partial y} = \frac{\partial}{\partial y}(\tan^{-1}\frac{y}{x} + k\pi) = \frac{1}{1 + (y/x)^2} \cdot \frac{1}{x} = \frac{x}{x^2 + y^2} = \frac{\partial u}{\partial x}$$

Hence the CR conditions are satisfied, and by **(3.7)**,

$$f'(z) = \frac{x}{x^2 + y^2} - \frac{iy}{x^2 + y^2} = \frac{x - iy}{x^2 + y^2} = \frac{1}{x + iy} = \frac{1}{z}$$

Note that $\tan^{-1} y/x + k\pi$ (or θ) is not continuous on the negative real axis, so the partial derivatives do not exist there. (The concept of analytic continuation covered in Chapter 9 makes it possible to define $\ln z$ over different domains so that its derivative exists at all points in restricted regions of those domains.)

Sometimes it is easier to use the definition of the derivative in an equivalent form known as the *difference quotient*.

THEOREM 3.8 (Difference Quotient Form)

The derivative of a complex function $f(z)$ exists at z iff the following limit exists and

(3.8)
$$\lim_{w \to z} \frac{f(w) - f(z)}{w - z} = f'(z)$$

Proof: Define $h = w - z$. Then $w = z + h$ and

$$\frac{f(w) - f(z)}{w - z} = \frac{f(z + h) - f(z)}{h}$$

Thus if the limit **(3.8)** exists, then so does $\lim \dfrac{f(z + h) - f(z)}{h}$ as $h \to 0$, which equals $f'(z)$, and **(3.8)** follows. Conversely, suppose the derivative exists; then $\lim \dfrac{f(w) - f(z)}{w - z}$ exists as $w \to z$ (or as $h \to 0$) and it equals $f'(z)$, thus **(3.8)**. ⬟

EXAMPLE 8

Find the derivative of the power function z^n for all integers $n > 0$ and show that it equals nz^{n-1}.

SOLUTION
We remind the reader of these factoring laws in $\mathbb{R}$:

$$x^2 - y^2 = (x - y)(x + y) \qquad \text{(2 terms in second parentheses)}$$
$$x^3 - y^3 = (x - y)(x^2 + xy + y^2) \qquad \text{(3 terms)}$$
$$x^4 - y^4 = (x - y)(x^3 + x^2 y + xy^2 + y^3) \qquad \text{(4 terms)}$$
$$\vdots$$
$$x^n - y^n = (x - y)(x^{n-1} + x^{n-2} y + x^{n-3} y^2 + \cdots + y^{n-1}) \qquad (n \text{ terms})$$

These same laws also apply in $\mathbb{C}$ since the algebra for $\mathbb{C}$ and $\mathbb{R}$ is the same. Using **(3.8)** we must find the limit

$$\lim_{w \to z} \frac{w^n - z^n}{w - z}$$

The factoring rule mentioned above gives us

$$\lim_{w \to z} \frac{(w - z)(w^{n-1} + w^{n-2} z + w^{n-3} z^2 + \cdots + wz^{n-2} + z^{n-1})}{w - z}$$

$$= \lim_{w \to z} (w^{n-1} + w^{n-2}z + w^{n-3}z^2 + \cdots + wz^{n-2} + z^{n-1}) \qquad [n \text{ terms}]$$

$$= z^{n-1} + z^{n-2}z + z^{n-3}z^2 + \cdots + z \cdot z^{n-2} + z^{n-1} \qquad [n \text{ terms}]$$

$$= nz^{n-1}$$

NOTE: The power rule when n is an arbitrary rational number can be obtained using the chain rule (established in the next section) as in calculus. See Problem 18. ⟍

This section will close with the statement of a theorem analogous to a familiar theorem of calculus, whose proof in complex variables is identical to that of the real variable case.

THEOREM 3.9

If $f(z)$ is differentiable at z_0, then $f(z)$ is continuous at z_0.

FORMAL RULES OF DIFFERENTIATION

The familiar sum and product rules for differentiation can be established for complex functions using the same methods of proof that are used in calculus for real-valued functions. These and other rules will be briefly explored here to show similarity. Assume that $f(z)$ and $g(z)$ are differentiable. Then:

(3.10) Linear Sum Rule $\qquad \dfrac{d}{dz}[af(z) + bg(z)] = af'(z) + bf'(z) \qquad$ (a, b constants)

(3.11) Product Rule $\qquad \dfrac{d}{dz}[f(z)g(z)] = f'(z)g(z) + f(z)g'(z)$

(3.12) Quotient Rule $\qquad \dfrac{d}{dz}\left[\dfrac{f(z)}{g(z)}\right] = \dfrac{f'(z)g(z) - f(z)g'(z)}{[g(z)]^2} \qquad$ (where $g(z) \neq 0$)

(3.13) Chain Rule $\quad$ If $F(z)$ is the *composition* or *product* $f \circ g(z) \equiv f[g(z)]$ of two differentiable functions f and g, then the derivative of $F(z)$ exists, and $F'(z) = f'[g(z)]g'(z)$.

(3.14) L'Hospital's Rule $\quad$ Suppose that $f(z) \to 0$ and $g(z) \to 0$ as $z \to c$. If the derivatives of f and g exist and are continuous at $z = c$, and if $\lim g'(z) \neq 0$, then

$$\lim_{z \to c} \frac{f(z)}{g(z)} = \lim_{z \to c} \frac{f'(z)}{g'(z)}$$

The proofs of the first three rules are identical to those of the calculus. However, the chain rule requires an explicit proof in the complex case, as well as L'Hospital's rule. As in the case of real variables, the proof of the chain rule can be established directly from the approximation theorem. If we assume that $f'(w)$ exists, then **(3.6)** takes on the form for any point w:

(3.15) $\qquad\qquad f(w + k) - f(w) = f'(w)k + \varepsilon k \qquad (\varepsilon \to 0 \text{ as } k \to 0)$

Assume that the derivatives of f and g exist at the specific points $w = g(z)$ and z, respectively, where z and $g(z)$ are temporarily regarded as fixed. Also let $k = g(z + h) - g(z)$. As $h \to 0$, then $k \to 0$ (a consequence of the continuity of $g(z)$). Since $w + k = g(z + h)$, (3.15) implies

$$f[g(z + h)] - f[g(z)] = f'[g(z)][g(z + h) - g(z)] + \varepsilon[g(z + h) - g(z)]$$

Since $f[g(z + h)] = F(z + h)$ and $f[g(z)] = F(z)$ we obtain (by dividing both sides by h)

$$\frac{F(z+h) - F(z)}{h} = f'[g(z)]\frac{g(z+h) - g(z)}{h} + \varepsilon \cdot \frac{g(z+h) - g(z)}{h}$$

The limit as $h \to 0$ then yields the desired result:

$$F'(z) = f'[g(z)]g'(z) + 0 \cdot g'(z) = f'[g(z)]g'(z) \quad \backslash\!\backslash$$

NOTE: The advantage of this method of proof is that it applies to the case when the domain of g is real instead of complex. Thus, $w = g(t)$ can represent a *parametric curve*, where $g(t) = x(t) + iy(t)$, $a \le t \le b$. In this case, $F(t) = f[g(t)]$ and we take h real. If both $x(t)$ and $y(t)$ are differentiable, the steps in the above proof do not change, except for notation, and the final result is $F'(t) = f'[g(t)]g'(t)$. $\backslash\!\backslash$

The proof of (3.14) (L'Hospital's rule) proceeds as follows: By hypothesis, f and g are differentiable, therefore continuous at c, so $f(z) \to f(c) = 0$ and $g(z) \to g(c) = 0$ as $z \to c$. Also, we have

$$\lim_{z \to c} \frac{f(z) - f(c)}{z - c} = f'(c) \qquad \text{and} \qquad \lim_{z \to c} \frac{g(z) - g(c)}{z - c} = g'(c) \ne 0$$

By the quotient rule for limits, and since $f(c) = g(c) = 0$,

$$\lim_{z \to c} \frac{f(z)}{g(z)} = \lim_{z \to c} \frac{f(z) - f(c)}{g(z) - g(c)} = \lim_{z \to c} \frac{\dfrac{f(z) - f(c)}{z - c}}{\dfrac{g(z) - g(c)}{z - c}}\Bigg|_{z=c} = \frac{f'(c)}{g'(c)} = \lim_{z \to c} \frac{f'(z)}{g'(z)}$$

(since the derivatives of f and g are continuous at $z = c$). $\backslash\!\backslash$

NOTE: A significant contrast between real and complex variables involves the mean value theorem (MVT). In real variable calculus, this theorem is traditionally used in proving L'Hospital's rule. Recall that for real variables the MVT states that if $f(x)$ is continuous for $a \le x \le b$ and differentiable for $a < x < b$, then there exists c between a and b such that

(∗) $$f'(c) = \frac{f(b) - f(a)}{b - a}$$

However, in complex variables, there exist differentiable functions for which (∗) does not hold for any value of c. (See Problem 16.) $\backslash\!\backslash$

EXAMPLE 9

Use the chain rule to find $\dfrac{d}{dz}(z^2 - 5z)^3$.

SOLUTION

$$\frac{d}{dz}(z^2 - 5z)^3 = 3(z^2 - 5z)^2[(z^2 - 5z)'] = 3(z^2 - 5z)^2(2z - 5)$$

EXAMPLE 10

Using the theorems on limits, evaluate the following limit in two ways: (1) by factoring and simplifying, and (2) using L'Hospital's rule:

$$\lim_{z \to 2i} \frac{z^3 + 4z}{z^4 - 16}$$

SOLUTION

(1) $$\lim_{z \to 2i} \frac{z^3 + 4z}{z^4 - 16} = \lim_{z \to 2i} \frac{z(z^2 + 4)}{(z^2 + 4)(z^2 - 4)} = \lim_{z \to 2i} \frac{z}{z^2 - 4} = \frac{2i}{(2i)^2 - 4} = -\tfrac{1}{4}i$$

(2) $$\lim_{z \to 2i} \frac{z^3 + 4z}{z^4 - 16} = \lim_{z \to 2i} \frac{3z^2 + 4}{4z^3} = \frac{3(2i)^2 + 4}{4(2i)^3} = \frac{-12 + 4}{4(-8i)} = \frac{-8}{-32i} = \frac{1}{4i} = -\tfrac{1}{4}i$$

EXAMPLE 11

Use the approximation theorem to estimate the value of $(3.01 + 2.02i)^3$. Compare with the actual value.

SOLUTION

Let $h = .01 + .02i$. With z^3 as the function $f(z)$ at $z = 3 + 2i$, then $(z^3)' = 3z^2$. By the approximation theorem,

$$(z + h)^3 = z^3 + 3z^2h + \{\text{Error}\} \approx (3 + 2i)^3 + 3(3 + 2i)^2(.01 + .02i)$$

$$= 3^3 - 3 \cdot 3 \cdot 2^2 + i(3 \cdot 3^2 \cdot 2 - 2^3) + 3(9 + 12i - 4)(.01 + .02i)$$

$$= -9 + 46i + (15 + 36i)(.01 + .02i)$$

$$= -9.57 + 46.66i$$

The actual value is given by

$$(3.01 + 2.02i)^3 = (3.01)^3 - 3(3.01)(2.02)^2 + i[3(3.01)^2(2.02) - (2.02)^3]$$

$$= -9.57511 + 46.66180i$$

DERIVATIVES AND ANTI-DERIVATIVES OF THE BASIC FUNCTIONS

The basic functions defined in Chapter 2 are all differentiable, sometimes for all z in the complex plane. Their derivatives follow the same pattern as their real counterparts. For example, as you are to work out in detail in Problem 3 below,

$$\frac{d}{dz}(\sin z) = \cos z$$

As in real variable calculus, the corresponding anti-derivative is written

$$\int \cos z \, dz = \sin z + C$$

where C is an arbitrary constant. This formula can be referred to as an *indefinite integral*, as in ordinary calculus, where the traditional appendix dz is, as usual, merely a symbol, often left to the imagination. But since it is convenient as a device to indicate the variable of integration, we shall use it here without attaching too much meaning to it at the present time. Other formulas for the derivative and anti-derivative for the elementary functions are similar.

A short list of these formulas appears in Table 1 below, which can all be worked out routinely; the details will be left as problems. If we take for granted that the CR conditions hold, then there are three ways to obtain these formulas: by definition, by use of **(3.4)**, or by the difference quotient method **(3.8)**.

DERIVATIVES	ANTI-DERIVATIVES
$\dfrac{d}{dz} z^n = n z^{n-1}$	$\displaystyle\int z^n dz = \dfrac{z^{n+1}}{n+1} + C \quad (n \neq -1)$
$\dfrac{d}{dz} e^z = e^z$	$\displaystyle\int e^z dz = e^z + C$
$\dfrac{d}{dz} \sin z = \cos z$	$\displaystyle\int \sin z\, dz = -\cos z + C$
$\dfrac{d}{dz} \cos z = -\sin z$	$\displaystyle\int \cos z\, dz = \sin z + C$
$\dfrac{d}{dz} \tan z = \sec^2 z$	$\displaystyle\int \sec^2 z\, dz = \tan z + C$
$\dfrac{d}{dz} \sinh z = \cosh z$	$\displaystyle\int \sinh z\, dz = \cosh z + C$
$\dfrac{d}{dz} \cosh z = \sinh z$	$\displaystyle\int \cosh z\, dz = \sinh z + C$
$\dfrac{d}{dz} \tanh z = \operatorname{sech}^2 z$	$\displaystyle\int \operatorname{sech}^2 z\, dz = \tanh z + C$
$\dfrac{d}{dz} \ln z = \dfrac{1}{z} \quad (\mathscr{R}z > 0)$	$\displaystyle\int \dfrac{dz}{z} = \ln z + C$
$\dfrac{d}{dz} a^z = a^z \ln a \quad (a > 0)$	$\displaystyle\int a^z dz = \dfrac{a^z}{\ln a} + C \quad (a > 0)$

TABLE 1

Note that in reversing any derivative formula to obtain the anti-derivative, an arbitrary constant must be added for the *most general* anti-derivative, just as in real variables. Recall that in calculus, if $f'(x) = 0$ for all x, then $f(x) = C$ (constant), a result of the mean value theorem. In spite of the absence of a mean value theorem in complex variables, this is also true for $\mathbb{C}$ (see Problem 17).

EXAMPLE 12

Establish the following version of *integration by parts* by proving the formula

(3.16)
$$\int uv = \left(\int u\right)v - \int \left(\int u\right)v'$$

and use it to obtain the anti-derivative $\int z^2 \sin z\, dz$.

SOLUTION

The derivative of the left side of **(3.16)** equals uv; by the product rule, that of the right side equals (with $U = \int u$ and $U' = u$):

$$\frac{d}{dz}(Uv) - \frac{d}{dz}\int Uv' = U'v + Uv' - Uv' = U'v = uv$$

The two derivatives are equal, so they differ by a constant C, which for convenience we temporarily take as zero, giving us **(3.16)**. (The constant must be included in the final answer).

By repeated use of **(3.16)** one obtains, with $u = \sin z$ and $v = z^2$ for the first step,

$$\int z^2 \sin z\, dz = \left(\int \sin z\, dz\right) \cdot z^2 - \int \left(\int \sin z\, dz\right) \cdot (2z)dz$$

$$= -z^2 \cos z + 2\int (\cos z)\cdot(z)\, dz$$

$$= -z^2 \cos z + 2\left[\sin z \cdot z - \int (\sin z \cdot 1)\, dz\right]$$

$$= -z^2 \cos z + 2z \sin z + 2\cos z + C$$

EXAMPLE 13

Use **(3.16)** to find an anti-derivative for $\ln z$.

SOLUTION

Here, the trick is to let $u = 1$ and $v = \ln z$ (as in real variable calculus). Thus,

$$\int 1 \cdot \ln z\, dz = \left(\int 1 dz\right)\cdot \ln z - \int \left(\int 1 dz\right)\cdot (\ln z)'\, dz = z \ln z - \int z \cdot \frac{1}{z}\, dz = z \ln z - z + C$$

PROBLEMS

1. Find the derivatives
   (a) $(6z^3 + 5z^2 - 10z + 1)'$
   (b) $[z \ln(z^2 + z + 1)]'$
2. Show that the CR conditions are valid for $f(z) = e^z$, and use **(3.7)** to find $f'(z)$.
3. Show that the CR conditions are valid for $f(z) = \sin z$, and use **(3.7)** to find $f'(z)$.
4. Use the quotient rule to find the derivative of $z^n = 1/z^m$ where $n = -m$ is a negative integer (and $m > 0$).
5. Use the chain rule to prove that $(\ln z)' = 1/z$. [**Hint:** Let $w = \ln z$ (or $e^w = z$) so that $\frac{d}{dz}(e^w) = \frac{d}{dz}(z) = 1$.]
6. Use the quotient rule to find

   (a) $\dfrac{d}{dz}(\tan z) \equiv \dfrac{d}{dz}\left(\dfrac{\sin z}{\cos z}\right)$

   (b) $\dfrac{d}{dz}\left(\dfrac{z+1}{z+2}\right)$

7. Find $d/dz[(z-1)(z^3 + z^2 + z + 1)] \equiv d/dz(z^4 - 1)$ in two ways:
   (a) By the product rule
   (b) By direct calculation of $d/dz(z^4 - 1)$.

8. Use **(3.16)** to find $\int z \cos z \, dz$

9. Use **(3.16)** to find $\int z^3 e^z dz$.

10. Use **(3.16)** to find $\int z \ln z \, dz$.

11. Show that the CR conditions for $f(z) = \bar{z} \equiv x - iy$ fail for all z, hence the derivative of $\bar{z}$ does not exist.

12. Use the approximation theorem to estimate $\sqrt{3.3 + 4.2i}$. Compare with the actual value.

13. Prove the following approximation theorem for real two-variable functions, [that is, **(3.5)**]: If the partial derivatives of a 2-variable function $u = u(x, y)$ exist and are continuous in some neighborhood of (x, y), then for all h and k, there exist ε and δ such that

(3.17)
$$u(x + h, y + k) = u(x, y) + \frac{\partial u}{\partial x}h + \frac{\partial u}{\partial y}k + \varepsilon h + \delta k$$

Moreover, as h and $k \to 0$ then both ε and $\delta \to 0$. [**Hint:** Use the notation $u_x(x, y)$ and $v_y(x, y)$ for the partial derivatives $\partial u/\partial x$ and $\partial u/\partial y$ evaluated at (x, y), and apply the one-variable approximation theorem to show that for each h and k, $[u(x + h, y + k) - u(x, y + k)] + [u(x, y + k) - u(x, y)] = [u_x(x, y + k)h + \varepsilon' h] + [u_y(x, y)k + \delta k] = u_x(x, y)h + u_y(x, y)k + \varepsilon h + \delta k$, where ε', $\delta \to 0$ as h, $k \to 0$ and $\varepsilon = \varepsilon' + u_x(x, y + k) - u_x(x, y)$.]

14. The approximation theorem of the previous problem can be used to prove the following identity that evaluates the derivative of $u(t)$ for a two-variable (real) function $u(x, y)$ when $x = x(t)$ and $y = y(t)$ are differentiable functions of t. In effect, this is the *chain rule* for a two-variable function, which is normally covered in multivariable calculus. Complete the details of proof which follow (assuming the partial derivatives exist) and where $U(t)$ represents the composite function $u(x(t), y(t))$.

(3.18)
$$U'(t) \equiv \frac{du}{dt} = \frac{\partial u}{\partial x}\frac{dx}{dt} + \frac{\partial u}{\partial y}\frac{dy}{dt}$$

(1) By definition $U'(t) = \lim\limits_{\Delta t \to 0} \dfrac{U(t + \Delta t) - U(t)}{\Delta t}$. We seek a more explicit representation for the numerator.

(2) Let $h = x(t + \Delta t) - x(t)$ and $k = y(t + \Delta t) - y(t)$. Then by **(3.5)**,
$$U(t + \Delta t) = u[x(t + \Delta t), y(t + \Delta t)] = u[x(t) + h, y(t) + k] = U(t) + \frac{\partial u}{\partial x}h + \frac{\partial u}{\partial y}k + \varepsilon h + \delta k$$
for all h, k, and Δt, where $\varepsilon \to 0$ and $\delta \to 0$ as h and $k \to 0$.

(3) This equation may be transformed into the following by simple algebra (dividing by Δt):
$$\frac{U(t + \Delta t) - U(t)}{\Delta t} = \frac{\partial u}{\partial x}\frac{x(t + \Delta t) - x(t)}{\Delta t} + \frac{\partial u}{\partial y}\frac{y(t + \Delta t) - y(t)}{\Delta t} + \varepsilon\frac{x(t + \Delta t) - x(t)}{\Delta t} + \delta\frac{y(t + \Delta t) - y(t)}{\Delta t}$$

(4) Take the limit as $\Delta t \to 0$; as $\Delta t \to 0$, h and $k \to 0$, and therefore ε, $\delta \to 0$. The result is **(3.18)**.

15. In the proof of the approximation theorem for complex variables **(3.6)**, the limits of each of the quantities ε_1, ε_2, ε_1', and ε_2' as $h \to 0$ equals zero. Subsequently the quantities $\delta_1 = \varepsilon_1 + i\varepsilon_1'$, $\delta_2 = \varepsilon_2 + i\varepsilon_2'$, and $\delta h = \delta_1 h_1 + \delta_2 h_2$ were defined. Show that as $h \to 0$, then $\delta \to 0$.

16. Show that $f(z) = e^z$ is a counterexample for the mean value theorem (as proposed in the Note on page 38) where $a = 0$ and $b = 2\pi i$.

17. Show that if $F(z)$ is an anti-derivative of $f(z)$, then all other anti-derivatives must be of the form $F(z) + C$ for some constant C. [**Hint:** Using **(3.18)**, show that if in some region $\partial u/\partial x \equiv 0$ and $\partial u/\partial y \equiv 0$, then u is constant.]

18. Use the chain rule to extend the power rule to non-rational n for the derivative of z^n ($z \neq 0$).

19. Prove the anti-derivative formulas by differentiation:

(a) $\int \sec^2 z \, dz = \tan z + C$

(b) $\int \sec z \tan z \, dz = \sec z + C$

(c) $\int \sec z \, dz = \ln(\tan z + \sec z) + C$ ($\sin z \neq 1$)

20. Use the results of Problem 19 and integration by parts **(3.16)** to find an anti-derivative for the function $f(z) = \sec^3 z$.

21. Evaluate $\lim\limits_{z \to 4} \dfrac{\sqrt{z} - 2}{z - 4}$ in two different ways: L'Hospital's rule and factoring as suggested by Example 2.

22. Evaluate $\lim\limits_{z \to 8} \dfrac{\sqrt[3]{z} - 2}{z - 8}$ in two different ways: L'Hospital's rule and factoring as suggested by Example 2.

23. Use L'Hospital's rule to evaluate

(a) $\lim\limits_{z \to 0} \dfrac{\sin z}{z}$

(b) $\lim\limits_{z \to 0} \dfrac{z - \sin z}{z^3}$

24. Show that the logarithm function is not continuous on the negative real axis, and thus is not differentiable there.

25. Prove that $|z|^2 \equiv x^2 + y^2$ is differentiable but not analytic at $z = 0$.

26. Show that the CR conditions are satisfied for the function

$$f(z) = \cos x \cosh y - i \sin x \sinh y$$

for all z ($= x + iy$) and find its derivative.

27. Suppose that $f(z) = u + i(2xy + y)$ for some function $u = u(x, y)$ and that $f(z)$ is analytic. Find u. Is u unique?

28. Show that if $f(z) = u(x, y) + iv(x, y)$ is analytic, then u satisfies Laplace's partial differential equation

$$\frac{\partial^2 u}{\partial x^2} + \frac{\partial^2 u}{\partial y^2} = 0$$

Important in engineering applications, such solutions to these p.d.e.'s are called **harmonic**.

29. Some Real Analysis for Mathematics Majors.   Provide the details of proof in the steps below for the theorem known as the **Bolzano-Weierstrass theorem** for real analysis, which states that if $x_1, x_2, x_3, \cdots$, $x_k, \cdots$ is any infinite sequence of real numbers defined on a closed interval $[a, b]$, then some subsequence $\{x'_n\}$ of $\{x_n\}$ converges to a point c on that interval.

(1) Without loss of generality, we may assume that $a = 0$ and $b > 0$. That is, if we prove the theorem for this case, then the result will follow for arbitrary intervals by simple translation.

(2) Choose the least integer m that is greater than b, and consider the following points on $[0, m]$

$$0, 1, 2, 3, 4, \cdots, m$$

These points define m subintervals of length 1 containing $[0, b]$, at least one of which, say $[b_0, b_0 + 1]$, contains an infinite number of x_n's.

(3) Consider, via decimals, the following points on $[b_0, b_0 + 1]$:

$$b_0.0, \ b_0.1, \ b_0.2, \ b_0.3, \ b_0.4, \ \cdots, \ b_0.9$$

These points divide the interval $[b_0, b_0 + 1]$ into 10 subintervals each of length $\frac{1}{10}$, at least one of which, say $[b_0.b_1, \ b_0.b_1 + \frac{1}{10}]$, contains an infinite number of x_n's.

(4) Consider the following points on $[b_0.b_1, \ b_0.b_1 + \frac{1}{10}]$:

$$b_0.b_10, \ b_0.b_11, \ b_0.b_12, \ b_0.b_13, \ b_0.b_14, \ \cdots, \ b_0.b_19$$

These points divide the interval $[b_0.b_1, \ b_0.b_1 + \frac{1}{10}]$ into 10 subintervals each of length $\frac{1}{100}$ at least one of which, say $[b_0.b_1b_2, \ b_0.b_1b_2 + \frac{1}{100}]$, contains an infinite number of x_n's.

(5) Continue in this manner, obtaining points $b_0.b_1b_2b_3 \cdots b_m$ such that $[b_0.b_1b_2b_3 \cdots b_m, \ b_0.b_1b_2b_3 \cdots b_m + 10^{-m}]$ contains infinitely many x_n's and is of length $1/10^m$ for each m. Choose a member of the sequence x_n that lies in the mth subinterval $[b_0.b_1b_2b_3 \cdots b_m, \ b_0.b_1b_2b_3 \cdots b_m + 10^{-m}]$. Re-label this member x_m.

(6) Consider the real number $c = b_0.b_1b_2b_3 \cdots b_m b_{m+1} \cdots$ and the subsequence $\{x_m\}$. One obtains $\lim x_m = c$ as $m \to \infty$. Also, since $x_m \leq b$ for all m, it follows that $0 \leq c \leq b$. [Note that the choices made for the digits b_m can cause c to be of the form

$$b_0.b_1b_2b_3 \cdots b_k 999 \cdots 9 \cdots$$

for some k, which can be shown to equal the terminating decimal $b_0.b_1b_2b_3 \cdots (b_k + 1)000 \cdots 0 \cdots$.]

30. **More Real Analysis for Mathematics Majors.** Extend the theorem of Problem 29 to include the 2-dimensional version in order to prove that every infinite sequence of points $\{P_n\}$ in a closed disk (the set of all points lying on or inside a circle of radius $r > 0$) has a subsequence converging to a point in that disk. [**Hint:** Let $P_n = (x_n, y_n)$ denote the given sequence of points, and suppose that (a, b) is the center of the disk. Then x_n lies on a closed interval $I = [a - r, a + r]$ for each n, and by Problem 29, some subsequence $\{x'_n\}$ converges to c on I; this subsequence defines a corresponding subsequence of points $P'_n = (x'_n, y'_n)$. Continue in like fashion for $\{y'_n\}$ to obtain a subsequence $\{y''_n\}$ of $\{y'_n\}$ that converges to d, and define $P''_n = (x''_n, y''_n)$. You must also show that the resulting point (c, d) lies in the disk.]

31. **Bolzano-Weierstrass Theorem for Complex Variables.** Prove the following version of the result of Problem 30:

Given a sequence of complex numbers $\{z_n\}$ lying in a closed disk D (all z such that $|z - z_0| \leq r$ for some z_0 and $r > 0$), there exists a subsequence $\{z'_n\}$ of $\{z_n\}$ converging to some point $c \in D$.

INTEGRATION IN COMPLEX VARIABLES

The (definite) integral in real analysis is the limit of certain so-called *Riemann sums*. This is a complicated procedure, and it is most fortunate that one can easily evaluate a vast collection of these definite integrals using anti-derivatives and the fundamental theorem of calculus. That, of course, was one of the revolutionary ideas about calculus in its early days in the 1700s. The entire field of mathematical physics and engineering mathematics opened up and flourished as a result, and continues to be important today. What we want to do here is to continue to adjust familiar concepts to make them appropriate for complex variables, just as we did in the previous chapter for the derivative. A few surprises are in store for us, however, as we shall see.

A REVIEW OF THE DEFINITE INTEGRAL FOR REAL NUMBERS

Let's begin with an example. Recall how you evaluated the integral of x^2 on the interval [1, 4]:

$$\int_1^4 x^2 dx = \tfrac{1}{3} x^3 \Big|_1^4 = \tfrac{1}{3}(4^3 - 1^3) = 21$$

As you know, $\tfrac{1}{3} x^3$ is an anti-derivative of x^2, and the resulting answer 21 can be interpreted as the *area under the curve* $y = x^2$ between the vertical lines $x = 1$ and $x = 4$, the reason being that the definite integral was defined as the limit of Riemann sums which converges to this area. Figure 4.1 illustrates this for a function $f(x) > 0$ on the interval $[a, b]$. In general, a Riemann sum R_n is the sum of the areas of n rectangles with vertical sides passing through $a = x_0 < x_1 < x_2 < \cdots < x_n = b$, with heights $f(x_k^*)$ where x_k^* is a point on the interval $[x_{k-1}, x_k]$, and widths $\Delta x_k = x_k - x_{k-1}$ ($k = 1, 2, 3, \cdots, n$). (Taken together, the closed intervals $[x_{k-1}, x_k]$ consist of what is called a *partition* of the interval $[a, b]$.) Thus,

$$R_n = \sum_{k=1}^n f(x_k^*) \Delta x_k \qquad (\Delta x_k = x_k - x_{k-1})$$

Assuming all widths Δx_k approach zero as $n \to \infty$, the Riemann integral is defined as the limit

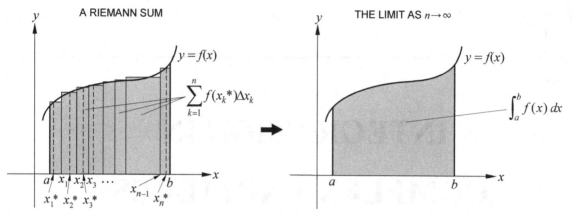

FIGURE 4.1

$$\int_a^b f(x)\,dx = \lim_{n\to\infty} R_n$$

provided the limit exists. (As pointed out in your calculus course, the integral does exist if $f(x)$ is continuous on $[a, b]$.) Moreover, this value is *independent of the Riemann sums R_n used to define it*—a somewhat amazing fact. Of course we can see intuitively, by geometry, that when $f(x) > 0$, the limit is the area under the curve between $x = a$ and $x = b$ regardless of the Riemann sums used. The function $f(x)$ is not required to be differentiable; if an anti-derivative can be found for $f(x)$, it can be used (as in the above example) to evaluate the integral, another amazing fact. These concepts are at the very heart of calculus. They are what give calculus the power to solve a variety of applied problems.

Recall the *fundamental theorem of calculus*. Since a similar theorem is true for complex variables, it is worthwhile to recall how it is presented in real variable calculus. It states that if $f(x)$ is any continuous function on the interval $[a, b]$, and if $F(x)$ is any anti-derivative of $f(x)$, then the Riemann integral of $f(x)$ over $[a, b]$ is given by the formula $\int_a^b f(x)\,dx = F(b) - F(a)$. This is how the answer 21 was obtained in the example at the beginning of this discussion. Now fast-forward to the subject at hand.

THE DEFINITE INTEGRAL IN COMPLEX VARIABLES

If we merely parrot the real analysis definition, the result is meaningless and indefinite. But the basic idea can be made valid, so let's look at some details. Let $f(z)$ be a complex function, and let c and d be any two complex numbers in the domain of $f(z)$. Suppose we select finitely many complex numbers in some order $c = z_0, z_1, z_2, z_3, \cdots, z_n = d$ in this domain. Define $\Delta z_k = z_{k-1} - z_k$, and choose z_k^* "between" z_{k-1} and z_k for each $k = 1, 2, 3, \cdots, n$. (To be more specific, one can take z_k^* as the midpoint between z_{k-1} and z_k.) Then, as in the real case, we want to consider the following limit, as each $\Delta z_k \to 0$:

(4.1) $$\int_c^d f(z)\,dz = \lim_{\substack{n\to\infty \\ \Delta z_k \to 0}} \sum_{k=1}^n f(z_k^*)\Delta z_k$$

But we are now dealing with the complex plane—a two-dimensional setting—and the points z_k can range over the entire plane. Thus, the above Riemann sums are not well defined. The remedy is to restrict the z_k's in some way. The most natural way is to use a curve C joining c and d and to require that the z_k's lie on C. For convenience, and because it includes virtually all curves, we use a **parametric representation** for curves in the xy-plane: $x = x(t), y = y(t), a \le t \le b$. In complex variables, this takes on the form $z = x(t) + iy(t)$ where $x(t)$ and $y(t)$ are real continuous functions, and t ranges over $[a, b]$. Thus,

$$z = z(t) \equiv x(t) + iy(t), \ a \le t \le b$$

We define a **curve** (sometimes also called a **contour**) to be a set of points given by such an equation. The curve is **continuous** provided the functions $x(t)$ and $y(t)$ are continuous, and it is called **smooth** if $x(t)$ and $y(t)$ are continuously differentiable (the derivatives exist and are continuous), and $z'(t) \ne 0$ on $[a, b]$. The curve is called **piecewise smooth** if it can be broken up into a finite number of smooth curves, where we still require $x(t)$ and $y(t)$ to be continuous on $[a, b]$. This allows a curve to have so-called *corner points*. A triangle, square, or rectangle are simple examples of piecewise smooth curves (in this case, **closed**, like the circle). A piecewise smooth curve can be shown to have finite length (thus called **rectifiable**). A spiral that wraps around a point can have infinite length and is therefore not rectifiable; on the other hand, a polygonal path with a million sides is rectifiable. (If $z(t)$ is continuous and one-to-one on $[a, b]$, the curve is called an **arc**, as in topology.)

Thus we are seeking the integral of $f(z)$ from c to d along some piecewise smooth curve $C : z = z(t)$, $a \le t \le b$, joining c and d (with $z(a) = c$ and $z(b) = d$), as illustrated in Figure 4.2. One begins by partitioning the interval $[a, b]$ on the t-axis by points $a = t_0 < t_1 < t_2 < \cdots < t_n = b$. The points $z_k = z(t_k)$ are then defined on C, with $z_k^* = z(t_k^*)$ for some t_k^* on $[t_{k-1}, t_k]$. If **(4.1)** is modified in this manner, a decent definition is achieved, and it is the standard definition for the integral in complex analysis.

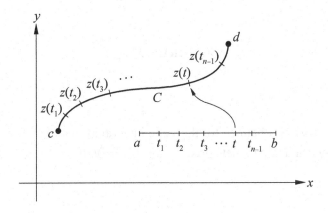

FIGURE 4.2

However, while this makes the definition **(4.1)** meaningful, it means that we are now dealing with so-called *line integrals*, integrals that are defined along curves. Different curves joining c and d can result in different integral values. You should keep this in mind as we go along. This is a characteristic feature of complex integration, unlike integrals of real-valued functions. An example will show how such line integrals are to be evaluated, anticipating a result that will be established later (Theorem 4.2 below). Such evaluations consist of simply substituting $x(t)$ and $y(t)$ into the integrand $[f(z) \equiv f(z(t))]$, replacing dz by $(dz/dt)dt \equiv z'(t)dt$, putting in the proper limits for t, and evaluating the resulting integral.

EXAMPLE 1

Suppose that $f(z) = \bar{z}$, $c = 0$, $d = 1 + i$, and that the curve C joining c and d is given by $z(t) = t^2 + t^3 i$, $0 \le t \le 1$. Here, $x = t^2$ and $y = t^3$ for $0 \le t \le 1$. Evaluate the integral **(4.1)** along C. (See Figure 4.3; C is part of a semi-cubical parabola.)

SOLUTION

First calculate dz/dt for C: $dz/dt = 2t + 3t^2 i$. Then one obtains

$$\int_c^d f(z)\,dz = \int_C \bar{z}\,dz = \int_0^1 \overline{t^2 + t^3 i}\,\frac{dz}{dt}\,dt = \int_0^1 (t^2 - t^3 i)(2t + 3t^2 i)\,dt$$

$$= \int_0^1 [(2t^3 + 3t^5) + i(3t^4 - 2t^4)]\,dt$$

$$= \int_0^1 [(2t^3 + 3t^5) + it^4]\,dt$$

$$= \left(\frac{t^4}{2} + \frac{t^6}{2} + \frac{it^5}{5} \right) \Bigg|_0^1 \quad \text{(using (4.3) and (4.4) below)}$$

$$= \frac{1}{2} + \frac{1}{2} + \frac{1}{5}i = 1 + \tfrac{1}{5}i$$

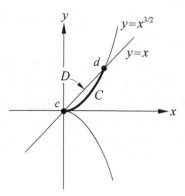

FIGURE 4.3

EXAMPLE 2

Show that a different answer is obtained for the same integral as that of Example 1 if evaluated along D, the straight line joining c and d, parameterized as $z(t) = (1 + i)t$, $0 \le t \le 1$, also shown in Figure 4.3).

SOLUTION
Here, $dz/dt = 1 + i$, so

$$\int_c^d f(z)\,dz = \int_D \bar{z}\,dz = \int_0^1 (1 - i)t \cdot (1 + i)\,dt = \int_0^1 2t\,dt = t^2 \Big|_0^1 = 1$$

In order to obtain formulas in general that are needed to evaluate a complex integral, it is necessary to assume that the curve of integration is smooth (that is, the parametric form $z(t) = x(t) + iy(t)$ of the curve is continuously differentiable). Under this assumption, certain manipulations with the Riemann sums in **(4.1)** will prove that the integral exists if $f(z)$ is continuous. The details appear in Problem 23.

THEOREM 4.2

If $f(z)$ is continuous, the integral **(4.1)** exists along any smooth curve $C : z = z(t)$, $a \le t \le b$, and is given by

(4.2)
$$\int_C f(z)\,dz = \int_a^b f(z(t))\,\frac{dz}{dt}\,dt$$

The following is another example showing the use of **(4.2)**.

EXAMPLE 3

Let C be the curve given by $z = t + ie^t$, $0 \leq t \leq 1$, as shown in Figure 4.4 (C lies on the graph of $y = e^x$, joining the points $z = i$ and $z = 1 + ei$.) Evaluate the complex integral $\int_C e^z \, dz$.

SOLUTION

Sometimes a different parameterization of the curve will greatly simplify calculations. In this case, let t be replaced by $\ln s$ (that is, $s = e^t$), where s ranges from $e^0 = 1$ to $e^1 = e$. The curve C is then represented by $z = \ln s + is$, $1 \leq s \leq e$. By **(4.2)**,

$$\int_C e^z \, dz = \int_1^e e^{\ln s + is} \frac{dz}{ds} \, ds = \int_1^e e^{\ln s}(\cos s + i \sin s)\left(\frac{1}{s} + i\right) ds$$

$$= \int_1^e (s \cos s + is \sin s)\left(\frac{1}{s} + i\right) ds$$

$$= \int_1^e [(\cos s - s \sin s) + i(s \cos s + \sin s)] ds$$

$$= (s \cos s + is \sin s)\Big|_1^e = e \cos e - \cos 1 + (e \sin e - \sin 1)i$$

[Notice that this answer is equivalent to

$$e(\cos e + i \sin e) - (\cos 1 + i \sin 1) = e^{1+ei} - e^i = e^z \Big|_i^{1+ei} = \int_i^{1+ei} e^z \, dz$$

which is formally what one would expect from ordinary calculus. We prove a theorem about this later.]

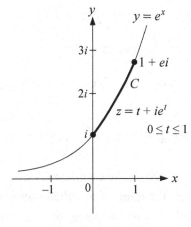

FIGURE 4.4

At this point, one might wonder whether the particular parameterization of the curve C matters. That is, does the value of the integral depend on the parameterization? Problem 24 below addresses this issue.

PROPERTIES OF THE INTEGRAL

One naturally expects the basic properties that are valid for real integrals to hold for complex integrals. If one uses Riemann sums and properties of limits, the proofs of these properties are straightforward.

(4.3) Constant Multiple Theorem $\int_C cf(z) \, dz = c \int_C f(z) \, dz$ (c = constant)

(4.4) Sum Theorem $\int_C [f(z) + g(z)]\,dz = \int_C f(z)\,dz + \int_C g(z)\,dz$

(4.5) Change of Variable   Let $t = g(s)$ be a real differentiable function. Then

$$\int_{g(a)}^{g(b)} f[z(t)]\frac{dz}{dt}\,dt = \int_a^b f[z(g(s))]\frac{dz}{ds}\,ds$$

Two further properties will be very important for later work. We let $-C$ denote the curve C traversed in the reverse direction; that is, if $C = \{z = z(t),\ a \le t \le b\}$ then $-C$ is defined by $z = z(s),\ a \le s \le b$ where $s = -t + a + b$. Also, for convenience we denote a curve that consists of two pieces, C_1 and C_2 (defined by separate parameterizations), by $C_1 + C_2$. Then

(4.6) Reverse Direction Theorem $\int_{-C} f(z)\,dz = -\int_C f(z)\,dz$

(4.7) Sum of Two Curves Theorem $\int_{C_1 + C_2} f(z)\,dz = \int_{C_1} f(z)\,dz + \int_{C_2} f(z)\,dz$

The proofs of these properties are strictly routine. As an example, the proof of the sum property **(4.4)** proceeds as follows: Assume that the integrals exist along some curve C; then in terms of Riemann sums (using algebra and the sum theorem on limits),

$$\int_C [f(z) + g(z)]\,dz = \lim_{n\to\infty} \sum_{k=1}^n [f(z_k) + g(z_k)]\Delta z_k$$

$$= \lim_{n\to\infty} \left\{ \sum_{k=1}^n f(z_k)\Delta z_k + \sum_{k=1}^n g(z_k)\Delta z_k \right\}$$

$$= \lim_{n\to\infty} \sum_{k=1}^n f(z_k)\Delta z_k + \lim_{n\to\infty} \sum_{k=1}^n g(z_k)\Delta z_k$$

$$= \int_C f(z)\,dz + \int_C g(z)\,dz$$

To prove **(4.7)**, let $C = C_1 + C_2$, where C_1 and C_2 are given by $z = z(t)$, $a \le t \le c$ and $c \le t \le b$ respectively, for $a < c < b$. Then it follows that

$$\int_{C_1 + C_2} f(z)\,dz = \int_a^b f[z(t)]z'(t)\,dt = \int_a^c f[z(t)]z'(t)\,dt + \int_c^b f[z(t)]z'(t)\,dt = \int_{C_1} f(z)\,dz + \int_{C_2} f(z)\,dz$$

NOTE: As in real variables, the function $f(z)$ need not be continuous on C in order for the integral on C to exist, and C need not be smooth. This raises the issue of so-called *improper integrals* for complex integration, which we will not cover here in detail. We will however often use the fact that the theory is valid when the curves involved are only piecewise smooth. In such cases, we simply sum the integrals along each smooth part of C in order to obtain the integral on C, just as improper integrals are handled in real variable calculus.   \

EXAMPLE 4

Assume that C is the "elbow path" from $1 + i$ to $2 + 3i$ given by $C_1 + C_2$ where $C_1 = \{z = t + i,\ 1 \le t \le 2\}$, and $C_2 = \{z = 2 + ti,\ 1 \le t \le 3\}$ (see Figure 4.5). Evaluate the integral $\int_C 2z\,dz$.

SOLUTION

From **(4.2)** and the sum theorem,

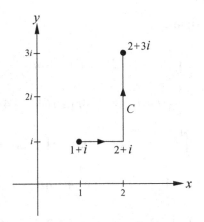

FIGURE 4.5

$$\int_C 2z\,dz = \int_{C_1} 2z\,dz + \int_{C_2} 2z\,dz$$

$$= \int_1^2 2(t+i)(1)dt + \int_1^3 2(2+ti)(i)dt$$

$$= \int_1^2 (2t+2i)dt + \int_1^3 (-2t+4i)dt$$

$$= (t^2+2ti)\Big|_1^2 + (-t^2+4ti)\Big|_1^3$$

$$= 4 + 4i - 1 - 2i + (-9 + 12i + 1 - 4i) = -5 + 10i$$

THE FUNDAMENTAL THEOREM FOR COMPLEX VARIABLES

The next example will illustrate an important point.

EXAMPLE 5

Evaluate the integral of Example 4 along the parabola $C = \{z = t + 1 + i(2t^2 + 1), 0 \le t \le 1\}$ (see Figure 4.6). (It follows that in this case C lies on the parabola $y = 2x^2 - 4x + 3$; to see this, set $x = t + 1$ and $y = 2t^2 + 1$, and substitute $x - 1$ for t.)

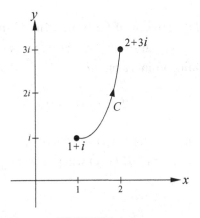

FIGURE 4.6

SOLUTION

Again using **(4.2)** and the basic principles,

$$\int_C 2z\,dz = 2\int_0^1 [t+1+i(2t^2+1)](1+4ti)dt$$

$$= 2\int_0^1 [(t+1-8t^3-4t)+i(4t^2+4t+2t^2+1)]dt$$

$$= \int_0^1 (-16t^3-6t+2)dt + i\int_0^1 (12t^2+8t+2)dt$$

$$= (-4t^4-3t^2+2t)\Big|_0^1 + i(4t^3+4t^2+2t)\Big|_0^1$$

$$= (-4-3+2)+i(4+4+2) = -5+10i$$

Note that we obtained the same answer for both Examples 4 and 5. Could it be that this integral is independent of path? There is something else to observe here.

NUMERICAL EXPERIMENT

(1) Calculate the values $(1+i)^2$ and $(2+3i)^2$.

(2) Find the difference $(2+3i)^2 - (1+i)^2$.

(3) Did anything happen in connection with Example 5?

The numerical experiment, together with the calculations made in Examples 4 and 5, show that,

$$\int_{1+i}^{2+3i} 2z\,dz = z^2\Big|_{1+i}^{2+3i}$$

One might wonder whether a "fundamental theorem of calculus" for complex integration is valid in general. Indeed:

THEOREM 4.8: Fundamental Theorem of Calculus for Complex Variables

Suppose that $f(z)$ is continuous and has an anti-derivative $F(z)$ that is valid in some region R. Then if C is any piecewise smooth curve in R joining points c and d,

$$\int_C f(z)\,dz = F(d) - F(c)$$

Proof: Let C be the curve given by $z = z(t) \equiv x(t) + iy(t)$, $a \le t \le b$, with $z(a) = c$ and $z(b) = d$. Further, suppose that $F(x+iy) = U(x,y) + iV(x,y)$ where $F'(z) = f(z)$. Since $F'(z)$ exists the CR conditions for F are valid in R:

$$\frac{\partial U}{\partial x} = \frac{\partial V}{\partial y} \quad \text{and} \quad \frac{\partial U}{\partial y} = -\frac{\partial V}{\partial x}$$

Also,

$$F'(z) = \frac{\partial U}{\partial x} + i\frac{\partial V}{\partial x} = f(z)$$

For convenience, define $U(t) = U(x(t), y(t))$ and $V(t) = V(x(t), U(t))$. Recall the calculus result (which was established in Problem 14, Chapter 3):

$$U'(t) = \frac{\partial U}{\partial x}\frac{dx}{dt} + \frac{\partial U}{\partial y}\frac{dy}{dt} \quad \text{and} \quad V'(t) = \frac{\partial V}{\partial x}\frac{dx}{dt} + \frac{\partial V}{\partial y}\frac{dy}{dt}$$

This requires the continuity of the partial derivatives of U and V, which follows since $f(z)$ [therefore $F'(z)$] is continuous. Thus by **(4.2)**, and using the CR conditions at the appropriate point,

$$\int_C f(z)\,dz = \int_a^b F'[(z(t)]\frac{dz}{dt}\,dt$$

$$= \int_a^b \left(\frac{\partial U}{\partial x} + i\frac{\partial V}{\partial x}\right)\left(\frac{dx}{dt} + i\frac{dy}{dt}\right)dt$$

$$= \int_a^b \left[\frac{\partial U}{\partial x}\frac{dx}{dt} - \frac{\partial V}{\partial x}\frac{dy}{dt} + i\left(\frac{\partial U}{\partial x}\frac{dy}{dt} + \frac{\partial V}{\partial x}\frac{dx}{dt}\right)\right]dt$$

$$= \int_a^b \left[\frac{\partial U}{\partial x}\frac{dx}{dt} + \frac{\partial U}{\partial y}\frac{dy}{dt}\right]dt + i\int_a^b \left[\frac{\partial V}{\partial y}\frac{dy}{dt} + \frac{\partial V}{\partial x}\frac{dx}{dt}\right]dt$$

$$= \int_a^b U'(t)\,dt + i\int_a^b V'(t)\,dt$$

$$= U(b) - U(a) + i[V(b) - V(a)]$$

$$= U(x(b), y(b)) + iV(x(b), y(b)) - U(x(a), y(a)) - iV(x(a), y(a))$$

$$= F[x(b) + iy(b)] - F[x(a) + iy(a)]$$

$$= F[z(b)] - F[z(a)] = F(d) - F(c) \quad \text{\\}$$

COROLLARY

If $f(z)$ is continuous and has an anti-derivative that is valid in some region R, then the integral of $f(z)$ from c to d is independent of path for all curves in R.

Proof: By the fundamental theorem, the final answer depends only on the two limits of integration (c and d) and the anti-derivative $F(z)$ of $f(z)$, and not on the curve C from c to d used to evaluate the integral. \\

NOTE: Since the integral of $f(z)$ is independent of path, the notation $\int_c^d f(z)\,dz$ becomes meaningful, and will be used from now on, when appropriate. \\

Theorem **4.8** guarantees that all the basic functions, those introduced in Chapter 3, have definite integrals that are independent of path (in certain connected regions). A converse for Theorem **4.8** is also true: if an integral of $f(z)$ is independent of path throughout an open region R, then $f(z)$ has an anti-derivative valid in R. (An **open** region is one for which every point has a neighborhood entirely enclosed by that region.) This result is important enough to state formally. Its proof is outlined in Problem 16.

THEOREM 4.9

If $\int_C f(z)dz$ is independent of path in some open region R, then $f(z)$ has an anti-derivative that is valid in R and the fundamental theorem holds. In particular, if one defines

$$F(z) = \int_c^z f(w)dw$$

along any curve joining c and z in R, then $F(z)$ is differentiable throughout R, and $F'(z) = f(z)$ for $z \in R$.

All of this theory has ramifications that are quite significant for complex analysis, which will be pursued in the next chapter. Right now, let's look at some numerical examples.

EXAMPLE 6

Evaluate the integral $\int_{1+3\pi i}^{2+2\pi i} e^z dz$.

SOLUTION

Since an anti-derivative of e^z is e^z itself, we obtain

$$\int_{1+3\pi i}^{2+2\pi i} e^z dz = e^z \Big|_{1+3\pi i}^{2+2\pi i} = e^2(\cos 2\pi + i\sin 2\pi) - e(\cos 3\pi + i\sin 3\pi) = e^2 + e$$

EXAMPLE 7

Evaluate the integral $\int_{-1-i}^{-1+i} \dfrac{dz}{z}$.

SOLUTION

$$\int_{-1-i}^{-1+i} \frac{dz}{z} = \ln z \Big|_{-1-i}^{-1+i} = \ln \sqrt{1^2+1^2} + 3\pi i/4 - \ln \sqrt{1^2+1^2} - (-3\pi i/4) = 3\pi i/2$$

EXAMPLE 8

Evaluate the integral $\int_C \dfrac{dz}{z}$ if C is the line segment from $-1-i$ to $-1+i$ without using the fundamental theorem. (Note that C can be parameterized by $z = -1 + ti, -1 \le t \le 1$.)

SOLUTION

$$\int_C \frac{dz}{z} = \int_{-1}^1 \frac{1}{-1+ti}(i)\,dt = \int_{-1}^1 \frac{i(-1-ti)}{1+t^2}\,dt = \int_{-1}^1 \frac{t-i}{t^2+1}\,dt$$

$$= [\tfrac{1}{2}\ln(t^2+1) - i\tan^{-1} t]\Big|_{-1}^1 = [\tfrac{1}{2}\ln 2 - \pi i/4] - [\tfrac{1}{2}\ln 2 + \pi i/4] = -\pi i/2$$

NOTE: The answers in the previous two examples do not agree, in apparent conflict with Theorem **4.8**. How can this be reconciled? [Answer: (You go first.) *The region R of validity for the fundamental theorem in this case does not contain the curve C.* In fact, any curve joining $-1-i$ and $-1+i$ that crosses the negative real axis is not in the domain of differentiability of $\ln z$ (see Example 7, Chapter3). The answer in Example 7 above is incorrect (the integral over $1/z$ is not independent of path and the fundamental theorem does not apply).] ⬊

CLOSED CURVES AND THE FUNDAMENTAL THEOREM

Up to now we have not explicitly considered closed curves for paths of integration. If a curve C is given by $z = z(t)$, $a \le t \le b$, it is said to be **closed** iff $z(a) = z(b)$. It is called **simple** provided it does not intersect itself [that is, $z(t_1) \ne z(t_2)$ for $t_1 \ne t_2$ on $[a, b]$, with the single exception $z(a) = z(b)$.] The theory of integration remains valid for closed curves. But a unique result pertinent to closed curves is very important.

THEOREM 4.10

In any region R, an integral on a simple closed curve in R equals zero iff it is independent of path in R.

Proof: Suppose $\int f(z)dz$ is independent of path. Let C be any simple closed curve in R, and choose $d \ne c = z(a)$ on C. Then $C = C_1 + C_2$, where C_1 is the part of C that extends from c to d and C_2 is the other part, from d back to c (see Figure 4.7). Since the integral is independent of path,

$$\int_{C_1} f(z)dz = \int_{-C_2} f(z)dz$$

hence by **(4.6)** and **(4.7)**,

$$\int_C f(z)cz = \int_{C_1} f(z)dz + \int_{C_2} f(z)dz = \int_{C_1} f(z)dz - \int_{-C_2} f(z)dz = 0$$

The converse follows in much the same manner. ⟍

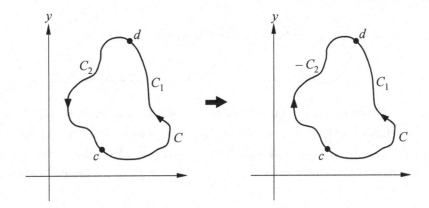

FIGURE 4.7

COROLLARY

If $f(z)$ has an anti-derivative that is valid throughout some region R, then $\int f(z)dz = 0$ over any closed curve completely contained in R.

EXAMPLE 9

Evaluate the following integrals on the unit circle $C : x^2 + y^2 = 1$ (that is, $|z| = 1$, parameterized by $z = e^{it}$, $0 \le t \le 2\pi$) without using the fundamental theorem.

(a) $\int_C z^2 dz$

(b) $\int_C \dfrac{dz}{z}$

SOLUTION

(a) $\int_C z^2 dz = \int_0^{2\pi} e^{2ti} \cdot ie^{ti} dt = i\int_0^{2\pi} e^{3ti} dt = i\int_0^{2\pi} (\cos 3t + i\sin 3t) dt = (\frac{1}{3}\sin 3t - \frac{1}{3}\cos 3t)\Big|_0^{2\pi} = 0.$

(b) $\int_C \dfrac{dz}{z} = \int_0^{2\pi} \dfrac{1}{e^{ti}}(ie^{ti})dt = \int_0^{2\pi} i\,dt = it\Big|_0^{2\pi} = 2\pi i$ (We must conclude that $\int \dfrac{dz}{z}$ is not independent

of path.)

ARC LENGTH AND THE MASS OF A WIRE

The arc length of a curve can be expressed as a complex integral:

(4.11)
$$l(C) = \int_C |dz| = \int_a^b \left|\frac{dz}{dt}\right| dt = \int_a^b \sqrt{(dx/dt)^2 + (dy/dt)^2}\, dt$$

where $dz/dt = dx/dt + i\,dy/dt$. This gives the arc length of the curve $C = \{z = z(t), a \le t \le b\}$. (You should recall this formula from your calculus course.) It follows that the arc length of any smooth curve exists, which can be easily extended to a piecewise smooth curve. An application is to find the total mass (weight) of a piece of wire in the shape of a curve $C : z = z(t)$, $a \le t \le b$, whose density $\delta(z)$ is a real-valued function of $z \in C$. This value is given by

(4.12)
$$W = \int_C \delta(z)|dz| = \int_a^b \delta[z(t)]\left|\frac{dz}{dt}\right| dt$$

EXAMPLE 10

Use **(4.12)** to find the total mass of a wire in the shape of the sine curve from $(0, 0)$ to $(\pi/2, 1)$ if the density is given by the relation $\mathcal{I}(z)\sqrt{1 - \mathcal{I}^2(z)}$. (Thus, the required curve is $C : z = t + i\sin t$, $0 \le t \le \pi/2$, and since $\mathcal{I}(z) = y$, $\delta(z(t)) = y\sqrt{1 - y^2} = \sin t\sqrt{1 - \sin^2 t} = \sin t \cos t$. Note that the wire has zero density at the ends, and maximum density at its middle (when $t = \pi/4$), as illustrated in Figure 4.8.

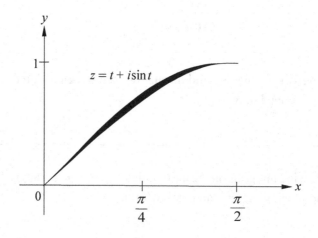

FIGURE 4.8

SOLUTION

By substitution into **(4.12)**, we obtain

$$W = \int_0^{\pi/2} \sin t \cos t \left| \frac{dz}{dt} \right| dt = \int_0^{\pi/2} \sin t \cos t \left| 1 + i \cos t \right| dt = \int_0^{\pi/2} \sin t \cos t \sqrt{1 + \cos^2 t} \, dt$$

$$= -\frac{1}{3}(1 + \cos^2 t)^{3/2} \Big|_0^{\pi/2} = -\frac{1}{3}(1 + 0^2)^{3/2} + \frac{1}{3}(1 + 1^2)^{3/2} = \frac{1}{3}(2\sqrt{2} - 1) \approx 0.60948$$

We close this chapter by obtaining two important inequalities for complex integrals that involve arc length. Suppose that C is parameterized by $z = z(t)$, $a \leq t \leq b$. The traditional notation is defined by

$$\int_C |f(z)||dz| = \int_a^b |f[z(t)]| \left| \frac{dz}{dt} \right| dt = \int_a^b |f[z(t)]| \sqrt{(dx/dt)^2 + (dy/dt)^2} \, dt$$

Now if $g(z)$ and $h(z)$ are continuous functions on C such that for all $z \in C$, $|g(z)| \leq |h(z)|$, then, since we are dealing with real integrals,

$$(*) \quad \int_C |g(z)f(z)||dz| = \int_a^b |g[z(t)]| |f[z(t)]| \left| \frac{dz}{dt} \right| dt \leq \int_a^b |h[z(t)]| |f[z(t)]| \left| \frac{dz}{dt} \right| dt = \int_C |h(z)||f(z)||dz|$$

It is also important to observe that the absolute value of the integral of f is less than or equal to the integral of the absolute value of f: using Riemann sums,

$$\left| \int_C f(z)\,dz \right| = \lim_{n \to \infty} \left| \sum_{k=1}^n f(z_k^*)\Delta z_k \right| \leq \lim_{n \to \infty} \sum_{k=1}^n |f(z_k^*)| |\Delta z_k|$$

$$= \lim_{n \to \infty} \sum_{k=1}^n |f(z_k^*)| |z_k'(t_k)| \Delta t_k = \int_a^b |f(z(t)| |dz/dt| \, dt = \int_C |f(z)| |dz|$$

This shows that, using $(*)$,

(4.13) $$\left| \int_C g(z)f(z)\,dz \right| \leq \int_C |g(z)||f(z)||dz| \leq \int_C |h(z)||f(z)||dz|$$

where $|g(z)| \leq |h(z)|$ for $z \in C$. Note what happens if $h(z) = M$ (constant) and therefore $|g(z)| \leq M$ on C:

(4.14) $$\boxed{\left| \int_C g(z)f(z)\,dz \right| \leq M \int_C |f(z)||dz| \qquad (|g(z)| \leq M)}$$

Further, a special case of **(4.14)** occurs if $f(z) = 1$. Using **(4.11)**,

(4.15) $$\boxed{\left| \int_C g(z)\,dz \right| \leq M \int_C |dz| \leq M \cdot l(C) \qquad (|g(z)| \leq M)}$$

EXAMPLE 11

Use **(4.15)** to find a bound for $\left| \int_C \dfrac{z}{z+5} \, dz \right|$ where C is the spiral $r = \theta$ in polar coordinates, $0 \leq \theta \leq \pi$, as shown in Figure 4.9. This curve can be converted into the parametric form $z = z(t)$ by observing that $r = \theta$ is equivalent to $x = t\cos t$, $y = t\sin t$ for $0 \leq t \leq \pi$ $(t = \theta)$; thus, $z = te^{it}$. [The length of this spiral involves the integral $\int \sec^3 u \, du$ and can be computed precisely by methods of calculus. A 2-place decimal approximation is $l(C) \approx 6.11$.]

SOLUTION

Observe that by the triangle inequality

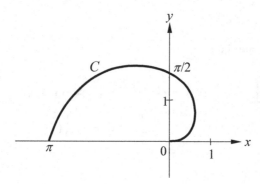

FIGURE 4.9

$$|z+5| \geq 5 - |z| \qquad (5 = |z+5-z| \leq |z+5| + |-z| = |z+5| + |z|)$$

By observing the graph in Figure 4.9, we can determine that the distance from 0 to z on C reaches its maximum at $z = \pi$ so. Thus $|z| \leq \pi$ and we obtain

$$\left| \frac{z}{z+5} \right| = \frac{|z|}{|z+5|} \leq \frac{|z|}{5-|z|} \leq \frac{\pi}{5-\pi} \qquad \text{for } z \in C$$

By (4.15),

$$\left| \int_C \frac{z}{z+5} dz \right| \leq \int_C \left| \frac{z}{z+5} \right| |dz| \leq \frac{\pi}{5-\pi} \int_C |\,dz\,| = \frac{\pi}{5-\pi} l(C) < 10.33$$

EXAMPLE 12

Use (4.14) to obtain a more refined estimate of the bound for the integral of Example 11.

SOLUTION

$$\left| \int_C \frac{z}{z+5} dz \right| \leq \int_C \left| \frac{1}{z+5} \right| |z| |dz| \leq \frac{1}{5-\pi} \int_C |z| |dz|$$

$$= \frac{1}{5-\pi} \int_0^\pi |\,te^{it}\,| \left| \frac{dz}{dt} \right| dt = \frac{1}{5-\pi} \int_0^\pi t \,|(1+ti)e^{it}| \,dt$$

$$= \frac{1}{5-\pi} \int_0^\pi t\sqrt{1+t^2}\, dt = \frac{1}{5-\pi} \cdot \frac{1}{3}(1+t^2)^{3/2} \Big|_0^\pi$$

$$= \frac{(1+\pi^2)^{3/2}-1}{3(5-\pi)} < 6.25$$

NOTE: The exact value for the integral of Example 9 can be found by the fundamental theorem since we can find an anti-derivative for the integrand: $z - 5\ln(z+5)$. Thus the integral has the value $\pi - 5\ln(\pi+5) + 5\ln 5 \approx 0.704$. ⧅

PROBLEMS

IN PROBLEMS 1–4, YOU ARE TO USE (4.2) TO EVLAUATE THE INTEGRAL $\int_C f(z)dz$, WHERE C IS A CURVE JOINING 0 AND $1+i$.

1. $f(z) = 4x + 2yi$, $C: \{z = t + ti \ (x = t \text{ and } y = t), \ 0 \leq t \leq 1\}$.

2. $f(z) = 4x + 2yi$, $C: \{z = t + t^2 i, \ 0 \leq t \leq 1\}$.

3. $f(z) = y - xi$, $C: \{z = t + t^3 i, \ 0 \leq t \leq 1\}$.

4. $f(z) = y - xi$, $C: \{z = t + t^4 i, \ 0 \leq t \leq 1\}$.

5. Evaluate $\int_C |z|^2 dz$ from 0 to $1 + i$ if

   (a) C is the straight line $y = x$,

   (b) C is the cubic curve $y = x^{3/2}$.

6. Evaluate the integral $\int_{-3i}^{6} z^2 dz$ by calculating $\int_C z^2 dz$ along the elbow path given by $C = C_1 + C_2$, where

   $C_1: \{z = t - 3i,\ 0 \le t \le 6\}$ and $C_2: \{z = 6 + 3ti,\ -1 \le t \le 0\}$.

7. Show that the answer to Problem 6 is the same as $\frac{1}{3}(6)^3 - \frac{1}{3}(-3i)^3$, thus showing that $\int_{-3i}^{6} z^2 dz = \frac{1}{3} z^3 \Big|_{-3i}^{6}$

8. Use **(4.2)** to evaluate $\int_C e^z dz$ where $C = \{z = ti,\ 0 \le t \le \pi\}$.

9. Evaluate the integral in Problem 8 using the fundamental theorem.

10. Use **(4.2)** to evaluate $\int_C \frac{1}{2}(e^{iz} + e^{-iz})\, dz$ where $C = \{z = ti,\ -2 \le t \le 1\}$.

11. Evaluate $\int_{-2i}^{i} \cos z\, dz$. (Compare with Problem 10.)

12. Evaluate $\int_{-2i}^{i} \sin z\, dz$ in two ways. [**Hint:** See Problems 10, and 11.]

13. Show that the fundamental theorem is valid for all curves C not crossing the negative x-axis and 0 for the integral

$$\int_C \frac{dz}{z}$$

14. Evaluate the following integral if C is the circle $|z - z_0| = 3$. [**Hint:** $z = z_0 + 3e^{it}$.]

$$\int_C \frac{dz}{z - z_0}$$

15. Evaluate the integral of Problem 14 if C is the circle $|z - z_0| = a$ (arbitrary a).

16. Provide the details of the following outline of the proof of Theorem **4.9**:

   (1) Let z be fixed, h arbitrary $\neq 0$, and let $\varepsilon > 0$, with $z + h$ in a neighborhood of z in R.

   (2) $\dfrac{F(z+h) - F(z)}{h} = \dfrac{1}{h}\left(\int_c^{z+h} f(w)\, dw - \int_c^{z} f(w)\, dw \right) = \dfrac{1}{h} \int_z^{z+h} f(w)\, dw$

   [**Hint:** Let C_1 be any curve from a to z in R and C_2 a curve from a to $z + h$ in R, with $C_2 = C_1 + C$, where C is the straight line segment joining z and $z + h$; use independence of path.]

   (3) Since $f(z)$ is constant with respect to w, $\int_z^{z+h} f(z)\, dw = hf(z)$.

   (4) $\left| \dfrac{F(z+h) - F(z)}{h} - f(z) \right| = \left| \dfrac{1}{h} \int_z^{z+h} [f(w) - f(z)]\, dw \right|$

   (5) Take h so close to zero that $|f(w) - f(z)| < \varepsilon$.

   (6) $\left| \dfrac{1}{h} \int_z^{z+h} [f(w) - f(z)]\, dw \right| \le \dfrac{1}{|h|} \int_z^{z+h} |f(w) - f(z)|\, |dw| < \varepsilon$. By (4), $\lim_{h \to 0} \dfrac{F(z+h) - F(z)}{h} = f(z)$.

17. Prove the property **(4.4)** for complex integrals.
18. Prove the property **(4.5)** for complex integrals.
19. Prove the property **(4.6)** for complex integrals.
20. Show that the arc length of $C: \{z = 6t + t^2 i,\ 0 \le t \le 4\}$ is $20 + 9\ln 3$. [From a table of integrals we find:
   $[\int \sqrt{a^2 + x^2}\, dx = \frac{1}{2} x\sqrt{a^2 + x^2} + \frac{1}{2} a^2 \ln(x + \sqrt{a^2 + x^2}).]$

21. Use **(4.14)** to find a bound for the integral $\int_C \dfrac{\sin z}{z + 2}\, dz$ where C is the curve of Problem 20.

22. Find the total mass of the wire $C: \{z = t^3 + ti,\ 0 \le t \le 1\}$ whose density function is $\delta(z) = \mathcal{R}(z) \equiv x$.

23. **Existence of the Complex Integral for a Continuous Function.** Complete the details in the following outline of the proof of Theorem **4.2**. (The method is to expand the terms in the Riemann sums of complex integrals into Riemann sums of real integrals, and to use the corresponding existence theorem for real integrals.)

(1) Consider for each index k the interval $[t_{k-1}, t_k]$ and the term $f(z_k^*)\Delta z_k$. Expand this term into its real and imaginary parts and show that, with $u(t)$ defined as $u(x(t), y(t))$ and $v(t) = v(x(t), y(t))$ for any t,

$$f(z(t_k^*))\Delta z_k = [u(t_k^*) + iv(t_k^*)][z(t_k) - z(t_{k-1})]$$

(2) To simplify, use u_k for $u(t_k^*)$, v_k for $v(t_k^*)$. The preceding then becomes

$$f(z(t_k^*))\Delta z_k = [u_k + iv_k][z(t_k) - z(t_{k-1})]$$
$$= [u_k + iv_k][x(t_k) + iy(t_k) - x(t_{k-1}) - iy(t_{k-1})]$$
$$= u_k \cdot [x(t_k) - x(t_{k-1})] - v_k \cdot [y(t_k) - y(t_{k-1})] + i\{u_k \cdot [y(t_k) - y(t_{k-1})] + v_k \cdot [x(t_k) - x(t_{k-1})]\}$$

(3) Since the derivatives $x'(t)$ and $y'(t)$ exist and are continuous, by the mean value theorem in $\mathbb{R}$, there exist points t_k' and t_k'' on (t_{k-1}, t_k) such that

$$x'(t_k') = \frac{x(t_k) - x(t_{k-1})}{t_k - t_{k-1}} \quad \text{and} \quad y'(t_k'') = \frac{y(t_k) - y(t_{k-1})}{t_k - t_{k-1}}$$

Thus, with $\Delta t_k = t_k - t_{k-1}$, we have $x(t_k) - x(t_{k-1}) = x'(t_k')\Delta t_k$ and $y(t_k) - y(t_{k-1}) = y'(t_k'')\Delta t_k$, and the above expression becomes

$$f(z(t_k^*))\Delta z_k = u_k x'(t_k')\Delta t_k - v_k y'(t_k'')\Delta t_k + iu_k y'(t_k'')\Delta t_k + iv_k x'(t_k')\Delta t$$

(4) Sum on k to obtain the sum of four Riemann-like sums for certain real integrals

$$\sum_{k=1}^{n} f(z(t_k^*))\Delta z_k = \sum_{k=1}^{n} u(t_k^*)x'(t_k')\Delta t_k - \sum_{k=1}^{n} v(t_k^*)y'(t_k'')\Delta t_k$$
$$+ i\sum_{k=1}^{n} u(t_k^*)y'(t_k'')\Delta t_k + i\sum_{k=1}^{n} v(t_k^*)x'(t_k')\Delta t_k$$

(5) A theorem from real analysis shows that if $g(t)$ and $h(t)$ are continuous and that if t_k' lies on $[t_k - t_{k-1}]$ for each k, then the pseudo-Riemann sum $\Sigma g(t_k^*)h(t_k')\Delta t_k$ converges to $\int g(t)h(t)dt$ as $n \to \infty$. It follows that the limits of the Riemann sums indicated above exist as $n \to \infty$, and by the sum of limits theorem,

$$\int_C f(z)\,dz = \int_a^b u(t)\frac{dx}{dt}\,dt - \int_a^b v(t)\frac{dy}{dt}\,dt + i\int_a^b u(t)\frac{dy}{dt}\,dt + i\int_a^b v(t)\frac{dx}{dt}\,dt$$

or, symbolically,

$$\int_C f(z)\,dz = \int_a^b [u(t) + iv(t)]\left(\frac{dx}{dt} + i\frac{dy}{dt}\right)dt$$

where $u(t) = u(x(t), y(t))$ and $v(t) = v(x(t), y(t))$. Since $f(z(t)) = u(t) + iv(t)$ and $dz/dt = dx/dt + idy/dt$, one obtains **(4.2)** above.

24. Prove that the integral along a curve C from c to d is independent of the particular (smooth) parameterization of C. That is, if C is given by both $z = z(t)$ $(a \le t \le b)$ and $w = w(s)$ $(a' \le s \le b')$, where dz/dt and dw/ds exist and are not zero, then

$$\int_a^b f[z(t)]\frac{dz}{dt}\,dt = \int_{a'}^{b'} f[w(t)]\frac{dw}{ds}\,ds$$

[**Hint:** Investigate a proof for **(4.5)** based on a change of variables for real integrals you studied in calculus.]

CAUCHY'S THEOREM AND

ITS APPLICATIONS

Further properties of the integral in complex variables are established in this chapter, including Cauchy's *theorem* and Cauchy's *integral formula*. These are among the most important concepts of complex analysis. They lead to two remarkable properties of complex functions, namely, (1) that if $f(z)$ is differentiable, then derivatives of all orders of $f(z)$ exist, and (2) if the values of an analytic function $f(z)$ are known on the *boundary* of a region, this completely determines its values inside. The statement of Cauchy's theorem itself requires us to define certain terms that involve a major area of mathematics, *topology*, as it applies to the complex plane. So the first section will be devoted to just that purpose.

EXTENSIONS OF PLANE GEOMETRY

In order to define precisely what is meant by a *region*, which up to now was a term used to informally describe the domain of a function, we present a few preliminary definitions. Some of these concepts may have been introduced to you in calculus.

First, recall the definition of *neighborhood*: A **neighborhood of a point** z_0 is the set $N(z_0, r)$ of all points z lying inside a disk of positive radius r centered at z_0, that is, the set of all z such that $|z - z_0| < r$ for $r > 0$. A set is **bounded** if it is contained by some neighborhood. Two sets are **disjoint** if they have no points in common.

A point z_0 is called an **interior point** of a set S provided S contains some neighborhood of z_0. If a neighborhood of z_0 contains no points of S, then z_0 is said to be an **exterior point** of S. If every neighborhood of z_0 contains both points in S and points in its **complement** (the set of all points not in S), then it is called a **boundary** or **limit point** of S (see Figure 5.1, left diagram). It follows that any point is either an interior point, an exterior point, or a boundary point of any set. It can be shown that every nonempty set except the whole plane always has at least one boundary point.

A set is said to be **closed** if it contains *all* its boundary points, and **open** if it contains *none* of its boundary points, as illustrated in Figure 5.1. It follows by definition that every point of a nonempty open set is an interior point. Consequently, an open set is a *union of neighborhoods*; that is, it is a *union of open disks* (sets of the form $|z - z_0| < r$ for $r > 0$). It can be shown that the complement of an open set is closed, and the complement of a closed set is open. (Can you prove this, based on the definitions?)

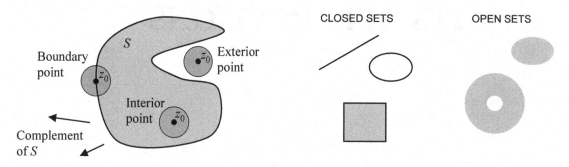

FIGURE 5.1

Finally, a **region** is a nonempty set R such that every neighborhood of every point of R contains at least one interior point of R. Thus, an open disk with some or all its boundary points is a region, while a circle or parabola by itself is not. (See Figure 5.2.) A region is said to be **connected** iff any two of its points can be joined by a polygonal path completely contained by that region, and **convex** iff any two of its points are the endpoints of a line segment completely contained by it. A bounded region is **simply-connected** iff both it and its complement are connected regions. (The intuitive notion of a simply-connected region R is that any closed curve in R can be "shrunk" to a point without leaving R.) A bounded region is **multiply-connected** iff it is connected and its complement is not connected; intuitively, a multiply-connected region is a region with one or more missing "holes". According to this definition, the **punctured** or **deleted disk** (a disk with its center point removed) is multiply connected. Finally, in order to eliminate pathology, the only regions we shall consider are bounded, and whose boundaries consist of one or more finitely many piece-wise smooth, simple closed curves (as defined in Chapter 4).

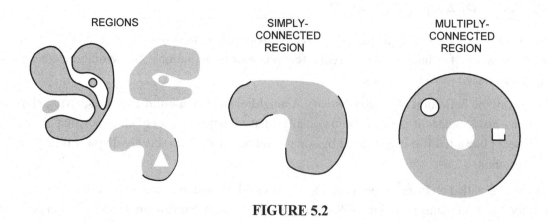

FIGURE 5.2

NOTE: The above definition of connectedness applies only to regions; an arc of a circle, for example, is certainly "connected", but no polygonal path joining any two of its points is contained by it. The more general definition states that a set is **connected** if and only if it is not made up of two or more subsets, called **components**, such that no one component meets the boundary of any other component. Another important concept needs to be mentioned. Essentially geometric, but involving much topology, a theorem known as the **Jordan closed curve theorem** states that any simple, closed curve C in the plane divides the plane into two disjoint open connected sets R and R', one called its "interior" or "inside"—the one that is bounded—the other, called its "exterior" or "outside". It readily follows that if R is the interior of a simple closed curve C, then R is simply-connected and the boundary of R is C. ⟍

The term *interior of a simple closed curve* is taken to mean the "inside" of C as described in the Jordan closed curve theorem, and not the topological *interior of a set*. Note also that the usage of the term *closed* as in "closed curve" has two possible meanings if taken out of context. Throughout, we rely on context to convey the intended meaning.

One last concept is needed: the *direction* of the path of integration for a closed curve. The counterclockwise direction is traditionally taken as the **positive** direction, while the clockwise direction is the **negative** direction. These concepts are, of course, intuitive, and they are difficult to make rigorous; they involve the very foundations of geometry. While this could be covered here, we choose not to do so since that would take us away from our intended goals. Without going into details, we regard *counterclockwise* on a simple-closed curve C to be that direction in which the interior of C stays to the "left" of a hypothetical person walking along the curve. Thus, concepts covered here and later will depend somewhat on an intuitive approach (as in virtually all treatments of complex variables).

CAUCHY'S THEOREM

Cauchy's theorem essentially says that if you integrate a function f on a simple closed curve C and f is analytic on C and in its interior, the answer is zero, regardless of how complicated the function or the curve of integration might be. This theorem has far-reaching implications in complex variables.

Of course, if $f(z)$ is a polynomial then the theorem is clearly true since such functions have antiderivatives whose integrals are thus independent of path (Theorem 4.10 and its corollary). In fact, for certain regions, the theorem is true likewise for all the elementary functions we introduced in Chapter 2. But the general statement given above is not that simple, and much of the literature reflects that. Many "short" proofs, such as those using Green's theorem (outlined in Problems 24–26 below for those who find it useful or necessary) require the continuity of the partial derivatives of u and v. The result known as the *Cauchy-Goursat theorem* assumes the mere existence of those partial derivatives and the CR conditions, and not their continuity. It was named after E. Goursat (1858–1936) who was the first to prove this. We give a proof of this stronger theorem here.

The proof we have chosen is adapted from one found in various sources, e.g., (Spiegel & Lipschutz, 2009, page 126). It requires an additional analysis of the behavior of a smooth curve in small neighborhoods. As in calculus, small neighborhoods of such curves are virtually straight lines. A well known result in curve theory (one that might seem obvious) is that if z_0 is some point of a piecewise smooth curve, and z is a variable point on C, the ratio of the length of the curve from z_0 to z to the length of the chord from z_0 to z, is approximately 1 (for z close to z_0). That is, if $l(z_0, z)$ denotes the arc length of C from z_0 to z,

(5.1)
$$\lim_{z \to z_0} \frac{l(z_0, z)}{|z_0 - z|} = 1$$

(See Problem 40.) It is not difficult to prove that if the neighborhood N of a point of a piecewise smooth curve C is sufficiently small, its intersection with C is a *connected* set.[†] Thus N meets C in a sub-curve C' of C. That is, $C' = N \cap C$ is the curve parameterized by $z = z(t)$, $a' \le t \le b'$ for some a', b' on $[a, b]$.

A special preliminary result establishes a bound for the length of C in small neighborhoods. For convenience, we shall often assume that the term *square* refers to the square as well as its interior, and similarly for the term *rectangle*.

LEMMA

Let z_0 be any point on a piecewise smooth curve C: $z = z(t)$, $a \le t \le b$. For all sufficiently small squares S containing z_0, S meets C in a sub-curve C' whose length is less than ¾ the perimeter of S.

Proof: Let $\varepsilon > 0$ be given. By **(5.1)** there exists a neighborhood N of z_0 such that $N \cap C$ is a sub-curve of C and for all $z \in N \cap C$
$$\left| \frac{l(z_0, z)}{|z_0 - z|} - 1 \right| < \varepsilon$$

(see Figure 5.3, left diagram). Multiply both sides by $|z_0 - z|$; since $|z_0 - z| \le l(z_0, z)$, the above inequality becomes
$$0 \le l(z_0, z) - |z_0 - z| < \varepsilon|z_0 - z| \qquad \text{or} \qquad l(z_0, z) < (1 + \varepsilon)|z_0 - z|$$

Let S be a square containing z_0, with side s and perimeter $p = 4s$, sufficiently small that S is contained by N and such that $S \cup C$ is a sub-curve C' of C (Figure 5.3, right diagram). If C_1 is the sub-curve of C' extending from $z_1 \in S$ to z_0, and C_2 is that from z_0 to $z_2 \in S$, then
$$l(C_1) < (1 + \varepsilon)|z_0 - z_1| \qquad \text{and} \qquad l(C_2) < (1 + \varepsilon)|z_0 - z_2|$$

Observe by geometry that $|z_0 - z_k| \le \sqrt{2}s$ for $k = 1, 2$. Hence
$$l(C') = l(C_1) + l(C_2) < (1 + \varepsilon)(|z_0 - z_1| + |z_0 - z_2|) \le (1 + \varepsilon)2\sqrt{2}s = (1 + \varepsilon)\sqrt{2}p/2$$

By taking $\varepsilon = 1/20$, we obtain $l(C') < \frac{3}{4}p$. ◊

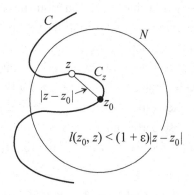
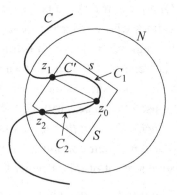

FIGURE 5.3

[†] If for all neighborhoods N of z_0, $N \cap C$ were not connected, and if T is the component containing z_0, there would exist a sequence of points z_n not in T converging to z_0, and C would not be a simple closed curve. This argument applies to neighborhoods which are squares instead of disks, and it also proves that $N \cap C$ has only a finite number of components for any neighborhood N.

THEOREM 5.2 Cauchy's Theorem for Simply-Connected Regions

If $f(z)$ is analytic on a simply-connected region R and on its boundary C, then

$$\int_C f(z)dz = 0$$

Proof: (1) Let S be a square containing $R \cup C$, and let p be its perimeter (Figure 5.4). Pass lines through the midpoints of each side of S, forming a subdivision of S into four congruent subsquares. These four sub-squares intersect $R \cup C$ in finitely many regions in groups of four or less, depending on the behavior of C. Disregard any subsquare that does not meet $R \cup C$. Orient the boundaries of each region in the counterclockwise direction (illustrated for a simple example in Figure 5.4, at right), and sum the integrals of $f(z)$ over these boundaries. The integrals over the common boundaries cancel each other since they occur in opposite directions [recall property **(4.6)**]. All that is left is the integral over C. Thus, if R_k represents the total boundary of the group of subregions inside the kth subsquare, then

$$\int_C f(z)dz = \int_{R_1} f(z)dz + \int_{R_2} f(z)dz + \int_{R_3} f(z)dz + \int_{R_4} f(z)dz$$

(there may be fewer than four main terms in this sum, and each $\int_{R_k} f(z)dz$ may consist of the sum of several sub-integrals). Now select from among the 4 or fewer integrals in this sum one having the largest magnitude, and re-label the corresponding group of regions R_1. Let the subsquare defining this choice be represented by S_1 (Figure 5.4 shows a typical selection.) We then obtain

$$\left| \int_C f(z)dz \right| = \left| \sum_{k=1}^{4} \int_{R_k} f(z)dz \right| \leq \sum_{k=1}^{4} \left| \int_{R_k} f(z)dz \right| \leq 4 \left| \int_{R_1} f(z)dz \right|$$

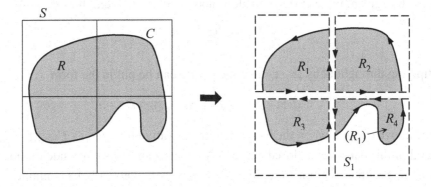

FIGURE 5.4

Apply this same process to R_1, subdividing the square S_1 into four congruent subsquares intersecting R_1, again forming four groups of regions. Orient their boundaries in the positive direction as before, sum the integrals of $f(z)$ over them, select the largest of the four or fewer integrals, and re-label the boundary of the selected group R_2, defined by the subsquare S_2. Thus, if R_2 also denotes the total boundary of $S_2 \cap R_2$, the second step of this procedure produces

$$\left| \int_{R_1} f(z)dz \right| \leq 4 \left| \int_{R_2} f(z)dz \right| \qquad \Rightarrow \qquad \left| \int_C f(z)dz \right| \leq 4^2 \left| \int_{R_2} f(z)dz \right|$$

Step three produces the group of regions R_3 with defining sub-square S_3 (see Figure 5.5 illustrating a choice for R_2 and R_3). One can observe that the perimeter of the sub-square S_n for

$n = 1, 2, 3$ is $p/2^n$. In general, the nth step produces the group of regions R_n (inside the subsquare S_n having perimeter $p/2^n$), with

(5.3)
$$\left| \int_C f(z)dz \right| \leq 4^n \left| \int_{R_n} f(z)dz \right|$$

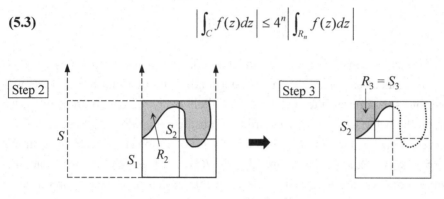

FIGURE 5.5

Note that if R_n is a full square, as shown for $n = 3$ in Figure 5.5 (right diagram), all the R_n's that follow will also be full squares. Otherwise, the boundary of R_n will contain a part C_n of C. (Keep in mind that the examples shown in Figures 5.4 and 5.5 are purely hypothetical.)

(2) For each n, select a point z_n in R_n (and on C_n if appropriate), forming a sequence of points in $R \cup C$. A subsequence of $\{z_n\}$ converges to some point z_0 in $R \cup C$ (Bozano-Weierstrass theorem). Since z_m lies in R_n for all $m \geq n$, it follows that z_0 belongs to R_n for all n (and to C if z_n lies in C_n from some point on). Now $f(z)$ is differentiable at z_0, so we have

$$\lim_{z \to z_0} \frac{f(z) - f(z_0)}{z - z_0} = f'(z_0)$$

Thus, if $\varepsilon > 0$ is given, there exists a neighborhood N of z_0 such that for all $z \in N$,

$$\left| \frac{f(z) - f(z_0)}{z - z_0} - f'(z_0) \right| < \varepsilon$$

By multiplying throughout by $|z - z_0|$ this inequality can be put in the form

$$|f(z) - f(z_0) - f'(z_0)(z - z_0)| < \varepsilon |z - z_0|$$

Notice the above expression on the left. An anti-derivative for $-f(z_0) - f'(z_0)(z - z_0)$ exists, so the integral of this part over a closed curve in N equals zero. Since the side of the sub-square S_n defining R_n becomes arbitrarily small as $n \to \infty$, S_n will eventually be contained by N and will satisfy the hypothesis of the lemma. It follows that [by **(4.14)**]

(*)
$$\left| \int_{R_n} f(z)dz \right| = \left| \int_{R_n} [f(z) - f(z_0) - f'(z_0)(z - z_0)]dz \right| \leq \varepsilon \int_{R_n} |z - z_0| \, |dz|$$

(3) We first obtain a bound for $|z - z_0|$ where $z \in R_n$. (See Figure 5.6 for the possibilities.)

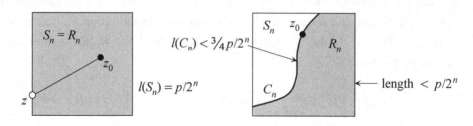

FIGURE 5.6

Since z and z_0 are contained by a square having perimeter $p/2^n$, $|z - z_0| < \sqrt{2}\,(\tfrac{1}{4}\,p/2^n) < \tfrac{1}{2}p/2^n$. Then, (∗) implies by **(4.15)**

(∗∗)
$$\left| \int_{R_n} f(z)dz \right| \le \frac{\tfrac{1}{2}p\varepsilon}{2^n} \cdot l(R_n)$$

(4) It remains to estimate the length of R_n. If R_n is a full square, then $l(R_n) = p/2^n$, as in Figure 5.6 (left diagram). Otherwise, R_n consists of a sub-curve C_n with length $< \tfrac{3}{4}p/2^n$ plus part of a full square with perimeter $p/2^n$, as in figure 5.6 (right diagram). In this case, $l(R_n) < \tfrac{3}{4}p/2^n + p/2^n < 2p/2^n$. Thus, in all cases, $l(R_n) < 2p/2^n$ and it follows from (∗∗) that

$$\left| \int_{R_n} f(z)dz \right| \le \frac{\tfrac{1}{2}p\varepsilon}{2^n} \cdot \frac{2p}{2^n} = \frac{p^2\varepsilon}{4^n}$$

If we return to the result **(5.3)**, we then obtain

$$\left| \int_{C} f(z)dz \right| \le 4^n \left| \int_{R_n} f(z)dz \right| \le 4^n \cdot \frac{p^2\varepsilon}{4^n} = p^2\varepsilon$$

Since ε was given arbitrarily, $\int_C f(z)dz = 0$. ◥

CAUCHY'S THEOREM FOR MULTIPLY-CONNECTED REGIONS

Cauchy's theorem can be extended to a multiply-connected region R, where if $f(z)$ is analytic on R and on its (disconnected) boundary C, then $\int_C f(z)dz = 0$. To establish this, start with order one, where the boundary of R consists of two curves C_1 and C_2, oriented as indicated in Figure 5.7 (diagram at far right). We shall prove that $\int_C f(z)dz = 0$, where $C = C_1 + C_2$. This is accomplished by a geometric "trick": draw a polygonal path D_1 in R from point α on C_1 to β on C_2 (shortened to a line segment in the figure), and similarly, a polygonal path D_2 in R from γ on C_2 to δ on C_1. This will form two simply-connected regions R_1 and R_2 bounded by the closed curves E_1 and E_2, both oriented in the counterclockwise direction. The path E_1 starts at α and goes along D_1 to β, then along C_2' (part of the curve C_2) from β to γ, along D_2 from γ to δ, then along C_1' (part of C_1) from δ back to α. Similarly, path E_2 goes along C_1'' (the other part of C_1) from α to δ, then along D_2, along C_2'' (the other part of C_2), and along D_1 back to α. The curves E_1 and E_2

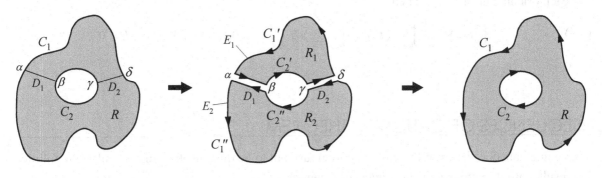

FIGURE 5.7

are closed curves bounding simply-connected regions R_1 and R_2, so the integrals of $f(z)$ over E_1 and E_2 are each zero by Theorem **5.2**. Thus $\int_{E_1} f(z)dz + \int_{E_2} f(z)dz = 0$ and since the integrals over D_1 and D_2 cancel,

$$0 = \int_{E_1} f(z)dz + \int_{E_2} f(z)dz = \int_{C_1'} f(z)dz + \int_{C_1''} f(z)dz + \int_{C_2'} f(z)dz + \int_{C_2''} f(z)dz$$

Therefore,

$$\int_{C_1} f(z)dz + \int_{C_2} f(z)dz = 0$$

NOTE: The reader will notice that orientation plays a critical role in the preceding equation. The proof depends on the fact that the direction on C_1 is opposite that of C_2. $\diagdown\hspace{-0.3em}\diagdown$

Define the **order** of a multiply-connected region to be the number of missing "holes", and suppose R is multiply connected of order $n > 1$ (such as that shown in Figure 5.8 for order 3), with boundary $C = C_1 + C_2 + C_3 + \cdots + C_n$ where the direction of C_1, the outer curve, is counterclockwise and the directions

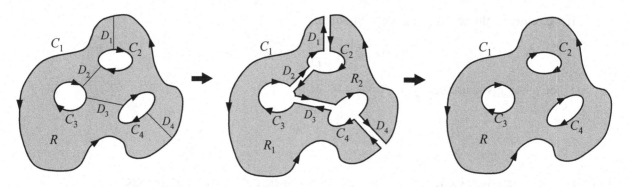

FIGURE 5.8

of $C_2, C_3, \cdots, C_n$ are each clockwise. As in the proof for order one, augment these curves with polygonal paths $D_1, D_2, D_3, \cdots, D_{n+1}$ again forming two simply-connected regions R_1 and R_2, about which the integrals of $f(z)$ vanish, where the integrals over these paths cancel. Thus, as before, the sum of the n integrals of $f(z)$ over $C_1, C_2, \cdots, C_n$ equals zero. (Figure 5.8 shows the details of this argument for $n = 3$.) This proves

THEOREM 5.4: Cauchy's Theorem for Multiply-Connected Regions

Suppose $f(z)$ is analytic on a multiply-connected region R of order n, and on its boundary C, where $C = C_0 + C_1 + \cdots + C_n$, with C_0 the outer boundary, oriented positively, and $C_1, C_2, \cdots, C_n$ the inner boundaries, each oriented negatively. Then

(5.4) $$\int_C f(z)dz \equiv \int_{C_0} f(z)dz + \int_{C_1} f(z)dz + \int_{C_2} f(z)dz + \cdots + \int_{C_n} f(z)dz = 0$$

CONSEQUENCES OF CAUCHY'S THEOREM

Two immediate corollaries of Cauchy's theorem turn out to be quite useful. In particular, the following result is used to prove *Cauchy's integral formula*.

COROLLARY A

Let $f(z)$ be analytic on R, where R is a multiply-connected region of order 1 having outer boundary C and inner boundary D, both taken counterclockwise (Figure 5.9). Then

(5.5)
$$\int_C f(z)dz = \int_D f(z)dz$$

Proof: Simply reverse the orientation of C_2 in **(5.3)** from clockwise to counterclockwise and re-label the curves, C for C_1 and D for C_2:

$$\int_C f(z)dz + \int_{-D} f(z)dz = 0$$

which is equivalent to

$$\int_C f(z)dz - \int_D f(z)dz = 0 \qquad \text{or} \qquad \int_C f(z)dz = \int_D f(z)dz \quad \backslash\!\backslash$$

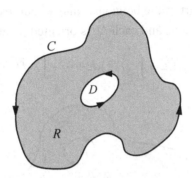

FIGURE 5.9

NOTE: The result **(5.5)** shows that $\int_C f(z)dz$ is independent of the boundary C and can be replaced by any curve D inside that encloses all the points where $f(z)$ is not analytic, if any. Thus we can sometimes simplify the integral by replacing C by a circle D inside C. The integral itself need not be zero. $\backslash\!\backslash$

EXAMPLE 1

Evaluate the integral $\int_C dz/z$ if

(a) C is the ellipse $\dfrac{(x-4)^2}{9} + \dfrac{y^2}{4} = 1$, and

(b) C is the ellipse $\dfrac{x^2}{9} + \dfrac{y^2}{4} = 1$ (see Figure 5.10).

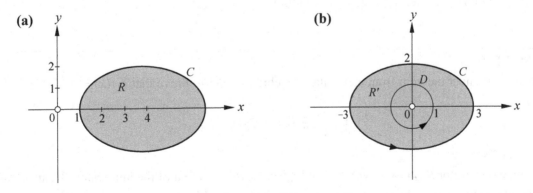

FIGURE 5.10

SOLUTION

(a) Since the ellipse does not contain $z = 0$, $1/z$ is analytic on C and in its interior R. By Cauchy's theorem $\int_C dz/z = 0$. (Direction of C is of no consequence here.)

(b) Take the direction of C to be positive, and let R' be its interior with $z = 0$ deleted. Let D be the unit circle $z = e^{it}$, $0 \le t \le 2\pi$ (taken positively). Then $1/z$ is analytic on R' and C. Thus by Corollary A,

$$\int_C \frac{dz}{z} = \int_D \frac{dz}{z} = \int_0^{2\pi} \frac{1}{e^{it}} \cdot \frac{dz}{dt}\, dt = \int_0^{2\pi} \frac{1}{e^{it}} (ie^{it})\, dt = i \int_0^{2\pi} dt = 2\pi i$$

COROLLARY B

Let R be a multiply-connected region of order n (illustrated in Figure 5.11 for $n = 3$) having boundary $C + D_1 + D_2 + \cdots + D_n$, with C the outer part of the boundary, and all curves oriented in the positive direction. If $f(z)$ is analytic on R and on its boundary, and each D_k is oriented positively, then

(5.6)
$$\int_C f(z)dz = \int_{D_1} f(z)dz + \int_{D_2} f(z)dz + \cdots + \int_{D_n} f(z)dz$$

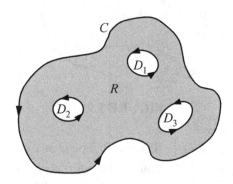

FIGURE 5.11

CAUCHY'S INTEGRAL FORMULA

Cauchy's formula takes on a rather unusual form, which takes getting used to. The closest we can come to it in real variables is illustrated by the following numerical example. (For the beginner, it is helpful to go over this before considering Cauchy's formula.)

EXAMPLE 2

Use the remainder theorem from elementary algebra in order to prove that if $f(x) = 3x^3 - 16x^2 + 11x + 7$, then

$$\int_3^5 \frac{f(x)}{x-2}\, dx = f(2)\ln 3$$

SOLUTION

Since the numerator $f(x)$ is a polynomial of higher degree than that of the numerator, the standard method of integration is to divide numerator by denominator. In this case, synthetic division can be applied:

$$
\begin{array}{r|rrrr}
2 & 3 & -16 & 11 & 7 \\
  &   & 6 & -20 & -18 \\
\hline
  & 3 & -10 & -9 & -11
\end{array}
$$

$$\int_3^5 \frac{3x^3 - 16x^2 + 11x + 7}{x - 2} dx = \int_3^5 \left(3x^2 - 10x - 9 + \frac{-11}{x - 2} \right) dx$$

$$= (x^3 - 5x^2 - 9x) \Big|_3^5 + \int_3^5 \frac{-11}{x - 2} dx$$

$$= 5^3 - 5 \cdot 5^2 - 45 - 27 + 45 + 27 + \int_3^5 \frac{-11}{x - 2} dx$$

$$= 0 + \int_3^5 \frac{-11}{x - 2} dx$$

$$= -11 \ln(x - 2) \Big|_3^5 = -11 \ln 3$$

By the remainder theorem of elementary algebra, $-11 = f(2)$. Hence

$$\int_3^5 \frac{f(x)}{x - 2} dx = f(2) \ln 3$$

A generalization of the formula obtained in Example 2 is easily shown: let $f(x)$ be any polynomial, and let $Q(x)$ be the polynomial quotient when $P(x)$ is divided by $x - c$. Then, if the integral of $Q(x)$ is zero,

(5.7) $$\int_a^b \frac{f(x)}{x - c} dx = f(c) \ln \left(\frac{b - c}{a - c} \right) \qquad (0 < c < a < b)$$

The complex variable version of the integral of **(5.7)** has a simplifying feature; one does not have to worry about the integral of the quotient $Q(z)$ being zero, because this follows automatically.

EXAMPLE 3

If $f(z) = z^4 + 5z^3 - 25z + 3$, evaluate the *complex integral*

$$\int_C \frac{f(z)}{z - 2} dz$$

where C is the circle of radius 1 centered at $z = 2$ (with parametric form $z = 2 + e^{ti}, 0 \le t \le 2\pi$).

SOLUTION

The remainder theorem is valid for complex variables since the rules of algebra are identical for both $\mathbb{R}$ and $\mathbb{C}$. Thus, by synthetic division one obtains $f(z)/(z - 2) = z^3 + 7z^2 + 14z + 3 + 9/(z - 2)$ with $f(2) = 9$ (the remainder theorem),

$$\int_C \frac{f(z)}{z - 2} dz = \int_C (z^3 + 7z^2 + 14z + 3) dz + \int_C \frac{f(2)}{z - 2} dz$$

The first integral on the right side in the above equation is zero since the integrand is a polynomial and is therefore analytic over the closed curve C and its interior. Thus

$$\int_C \frac{f(z)}{z - 2} dz = \int_C \frac{f(2)}{z - 2} dz = f(2) \int_0^{2\pi} \frac{1}{e^{it}} \left(\frac{dz}{dt} \right) dt$$

$$= f(2) \int_0^{2\pi} \frac{1}{e^{it}} \left(i e^{it} \right) dt = f(2) \int_0^{2\pi} i \, dt = 2\pi i f(2)$$

Note that the final result of the previous example can be written in the form

(5.8)
$$f(2) = \frac{1}{2\pi i} \int_C \frac{f(z)}{z-2} dz$$

This is a form of *Cauchy's formula* for the polynomial $f(z)$, which is true for any polynomial. But this formula is also true for *any analytic function $f(z)$*. In order to prove this, we shall need two results, which essentially allow us to *take limits and derivatives under the integral sign* (for special situations). In real variables, the relation

$$\lim_{y \to c} \int_a^b f(x,y) dx = \int_a^b \lim_{y \to c} f(x,y) dx$$

can be shown only under certain circumstances. You can experiment with the example

$$\lim_{y \to 0} \int_0^1 \frac{2xy}{(x^2+y^2)^2} dx$$

to discover for yourself some difficulties (see Problem 38).

The specific results needed are valid for any piecewise smooth closed curve C, with $f(z)$ analytic at all points on C and in its interior, and where $g(z) = z^n$ (n any integer):

(5.9)
$$\lim_{h \to 0} \int_C f(c+h) dz = \int_C f(c) dz \qquad (c \text{ constant})$$

and

(5.10)
$$\lim_{h \to 0} \int_C f(z) \left[\frac{g(z+h) - g(z)}{h} \right] dz = \int_C f(z) g'(z) dz \qquad [g(z) = z^n, z \neq 0]$$

For proofs, see Problems 41 and 42.

Now consider a function $f(z)$ that is analytic on a simply-connected region R and on its boundary C. Let c be an interior point of R, and let D_r be a circle of radius r centered at c, with D_r in R (as in Figure 5.12). Then, since $z - c = re^{it}$ for $0 \leq t \leq 2\pi$, and letting $h = re^{it}$, we obtain by **(5.5)**

$$\int_C \frac{f(z)}{z-c} dz = \int_{D_r} \frac{f(z)}{z-c} dz = \int_0^{2\pi} \frac{f(c+h)}{re^{it}} \left(rie^{it} \right) dt = i \int_0^{2\pi} f(c+h) dt$$

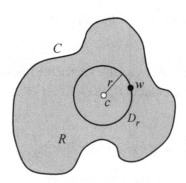

FIGURE 5.12

Now let $r \to 0$, and thus $h \to 0$. By **(5.9)**

$$\int_C \frac{f(z)}{z-c} dz = i \int_0^{2\pi} f(c) dt = f(c) \cdot i \int_0^{2\pi} dt = f(c) \cdot 2\pi i$$

from which is obtained the relation

(5.11) Cauchy's First Integral Formula $f(c) = \dfrac{1}{2\pi i} \displaystyle\int_C \frac{f(z)}{z-c} dz$

A rather startling fact implicit in **(5.11)** is that, in theory at least, if we know what $f(z)$ equals on the *boundary* of a region R, this determines the values $f(c)$ for any $c \in R$, thus for all z in the entire region R inside C. Moreover, a second formula similar to **(5.11)** can be derived, which will lead to an equally surprising result.

EXAMPLE 4

Use Cauchy's formula to evaluate the integral

$$\int_C \frac{\ln(z+2)}{z(z-2)} dz$$

where C is the unit circle with center $z = 2$.

SOLUTION

In this case, we use the set-up $c = 2$ and $f(z) = \ln(z+2)/z$. Then by **(5.11)**,

$$\int_C \frac{\ln(z+2)}{z(z-2)} dz = \int_C \frac{\ln(z+2)/z}{z-2} dz = 2\pi i \cdot \left. \frac{\ln(z+2)}{z} \right|_{z=2} = 2\pi i \cdot \frac{\ln 4}{2} = \pi i \ln 4$$

NOTE: It is important in these problems to make certain the hypothesis of Cauchy's formula is satisfied. In Example 4, $f(z)$ is analytic inside and on C, but not at $z = 0$, a point which, nevertheless, lies outside of C. ∖

EXAMPLE 5

Use Cauchy's formula to evaluate the integral of Example 4 if C is the unit circle centered at $z = 0$.

SOLUTION

This time, if $f(z) = \ln(z+2)/(z-2)$ then $f(z)$ is analytic for all $|z| \le 1$, hence inside and on C. Here, with $z_0 = 0$,

$$\int_C \frac{\ln(z+2)}{z(z-2)} dz = \int_C \frac{\ln(z+2)/(z-2)}{z} dz = 2\pi i \left. \frac{\ln(z+2)}{z-2} \right|_{z=0} = 2\pi i \cdot \frac{\ln 2}{-2} = -\pi i \ln 2$$

What if C encloses both points $z = 0$ and $z = 2$ for the integral considered in Examples 4 and 5? Then we would use **(5.6)**, choosing circles centered at these two points, small enough to lie inside C. The next example illustrates how this works.

EXAMPLE 6

Evaluate the integral

$$\int_C \frac{\ln(z+2)}{z(z-2)} dz$$

where C is the ellipse $\dfrac{x^2}{9} + \dfrac{y^2}{4} = 1$ of Example 1, pictured in Figure 5.10 (diagram at right).

SOLUTION

Here it is necessary to use circles D_1 and D_2 centered at $z = 0$ and $z = 2$ each having radius $\frac{3}{4}$ so they will lie in the interior of the ellipse. Then we have, by **(5.6)** and the results of Examples 4 and 5

$$\int_C \frac{\ln(z+2)}{z(z-2)}dz = \pi i \ln 4 - \pi i \ln 2 = \pi i \ln 2$$

CAUCHY'S INTEGRAL FORMULA FOR HIGHER ORDER

An interesting formula for the derivative $f'(z)$ at c has the same basic form as **(5.11)**. Note that by simple substitution, **(5.11)** yields for any complex number h close to zero

$$f(c+h) = \frac{1}{2\pi i}\int_C \frac{f(z)}{z-(c+h)}dz$$

If $h' = -h$, then by simple algebra and the use of the additive property of the integral,

$$f(c+h) - f(c) = \frac{1}{2\pi i}\int_C \left(\frac{f(z)}{z-c+h'} - \frac{f(z)}{z-c} \right) dz$$

and, dividing by $h = -h'$,

$$\frac{f(c+h)-f(c)}{h} = \frac{1}{2\pi i}\int_C f(z)(-1)\frac{(z-c+h')^{-1}-(z-c)^{-1}}{h'}dz$$

Note that by **(5.10)**, as $h' \to 0$ (thus $h \to 0$), the integral converges to

$$\frac{1}{2\pi i}\int_C f(z)(-1)\frac{d}{dz}(z-c)^{-1}dz = \frac{1}{2\pi i}\int_C f(z)(z-c)^{-2}dz$$

which proves that the expression on the left converges. Thus we obtain a formula for the *derivative* of $f(z)$ evaluated at c:

(5.12)
$$f'(c) = \frac{1}{2\pi i}\int_C \frac{f(z)}{(z-c)^2}dz$$

The same idea can be used to prove that $f''(c)$ exists and to obtain its formula. Applying the same algebra as above to **(5.12)** we have

$$\frac{f'(c+h)-f'(c)}{h} = \frac{1}{2\pi i}\int_C f(z)(-1)\frac{(z-c+h')^{-2}-(z-c)^{-2}}{h'}dz$$

In taking the limit as $h' \to 0$, by **(5.10)** the limits exist on both sides, and

(5.13)
$$f''(c) = \frac{1}{2\pi i}\int_C f(z)(-1)\frac{d}{dz}(z-c)^{-2}dz = \frac{2!}{2\pi i}\int_C \frac{f(z)}{(z-c)^3}dz$$

An important conclusion can be drawn that is a distinct departure from calculus [the result of **(5.13)**]. It will have far-reaching consequences, as we shall see in later chapters.

THEOREM 5.14

If a complex function $f(z)$ has a derivative $f'(z)$ on an open region R, then the second derivative $f''(z)$ exists on R, and $f(z)$ has derivatives of all orders on R.

The formulas **(5.12)** and **(5.13)** can be generalized: it can be proved in general that for any closed curve C having c in its interior, and for $f(z)$ any function that is analytic on C and in its interior,

> **(5.15) Cauchy's Integral Formula** $f^{(n)}(c) = \dfrac{n!}{2\pi i}\displaystyle\int_C \dfrac{f(z)}{(z-c)^{n+1}}\,dz$ (*n* any integer ≥ 0)

(In Problem 15 you are challenged to establish **(5.15)** by mathematical induction.)

EXAMPLE 7

Evaluate the integral $\displaystyle\int_C \dfrac{z\ln z}{(z-1)^6}\,dz$ where C is the circle $|z-1| = \frac{1}{2}$.

SOLUTION

Cauchy's formula **(5.15)** applies with $n = 5$ ($z\ln z$ is analytic for z not on the negative real axis or zero, thus for z on and inside C). This requires the 5th derivative of $z\ln z$:

$$(z\ln z)' = \ln z + 1, \quad (z\ln z)'' = 1/z, \quad (z\ln z)^{(3)} = -z^{-2}, \quad (z\ln z)^{(4)} = 2z^{-3}, \quad (z\ln z)^{(5)} = -6z^{-4}.$$

Then by **(5.15)**

$$\frac{5!}{2\pi i}\int_C \frac{z\ln z}{(z-1)^6}\,dz = (-6z^{-4})\Big|_{z=1} = -6 \qquad \text{or} \qquad \int_C \frac{z\ln z}{(z-1)^6}\,dz = \frac{(-6)\cdot 2\pi i}{5!} = -\frac{\pi i}{10}.$$

EXAMPLE 8

Show that if $P(z)$ is any monic polynomial of degree n (*monic* means that the coefficient of z^n is unity) and C is any simple closed curve surrounding $z = c$, then, regardless of the value of c,

$$\int_C \frac{P(z)}{(z-c)^{n+1}} = 2\pi i \qquad \text{and} \qquad \int_C \frac{P(z)}{(z-c)^m} = 0 \quad \text{for } m > n+1$$

SOLUTION

Note that the nth derivative of $P(z)$ is $n!$ But the $(n+1)$st derivative of $P(z)$ equals zero, as well as all derivatives of higher order. Using **(5.15)**,

$$\int_C \frac{P(z)}{(z-c)^{n+1}} = \frac{2\pi i}{n!}\cdot n! = 2\pi i \qquad \text{and} \qquad \int_C \frac{P(z)}{(z-c)^m} = \frac{2\pi i}{(m-1)!}\cdot 0 = 0 \quad (m > n+1)$$

APPLICATIONS: THE FUNDAMENTAL THEOREM OF ALGEBRA

An important inequality due to Cauchy will lead to a proof of what you have no doubt always taken for granted, namely, that a polynomial equation over the reals has at least one solution (either real or complex). This result is known as the *fundamental theorem of algebra*. It is interesting to learn how complex variables can be used to actually prove this basic result in a relatively simple manner. It involves several steps, to be undertaken next.

We begin with an analytic function defined on a circle C and its interior, with center c and radius r. It follows that $f(z)$ is bounded on C, and we can let M be a bound for $|f(z)|$. A simple use of Cauchy's formula will prove the following.

(5.16) Cauchy's Inequality   If $|f(z)| \leq M$ on the circle $|z - c| = r$, then $\left|f^{(n)}(c)\right| \leq \dfrac{Mn!}{r^n}$

Proof: By Cauchy's integral formula **(5.15)** and the inequality **(4.15)**, and letting C be the circle $|z - c| = r$,

$$\left|f^{(n)}(c)\right| = \left|\frac{n!}{2\pi i}\int_C \frac{f(z)}{(z-c)^{n+1}}\,dz\right| \leq \left|\frac{n!}{2\pi i}\int_C \frac{|f(z)|}{|z-c|^{n+1}}|\,dz\,|\right| \leq \frac{n!}{2\pi}\cdot\frac{M}{r^{n+1}}l(C) = \frac{n!M}{2\pi r^{n+1}}\cdot 2\pi r$$

THEOREM 5.17 (Liouville's Theorem)

If for all z in the entire complex plane $f(z)$ is both analytic and bounded, then $f(z)$ must be a constant.

Proof: Suppose $|f(z)| < M$ for all z. Then with $n = 1$ in Cauchy's inequality **(5.16)**,

$$|f'(c)| < \frac{M}{r}$$

for arbitrary complex c, and arbitrary r. Let $r \to \infty$. Then $|f'(c)| \leq 0$ or $f'(c) = 0$. That is, $f'(z) = 0$ for all z which implies that $f(z)$ is constant. ⦰

THEOREM 5.18: Fundamental Theorem of Algebra

Suppose that $P(z) = a_0 + a_1 z + a_2 z^2 + \cdots + a_n z^n$ is any polynomial over the complex numbers, with $n \geq 1$ and $a_n \neq 0$. Then there exists a point z_0 such that $P(z_0) = 0$.

Proof: Suppose that $P(z) \neq 0$ for all z. Then the function $f(z) = 1/P(z)$ is defined and analytic on the entire plane. Since $P(z)$ has degree ≥ 1, it is not constant, hence $f(z)$ is not constant, and by Louiville's theorem, $f(z)$ is unbounded. This means that for each positive integer n, we can find a point z_n such that $|f(z_n)| > n$. Suppose the sequence $\{z_n\}$ is unbounded. Then for each positive integer m, we can find a member of the sequence z_m such that $|z_m| > m$. Hence, $|z_m| \to \infty$ for these values of m. But, since $P(z)$ is a polynomial, $|P(z_m)| \to \infty$[†] and $|f(z_m)| \to 0$ as $m \to \infty$. But this contradicts $|f(z_n)| > n \geq 1$ for all n. Therefore, the points z_n must lie on a closed disk D in the complex plane. By the Bolzano-Weierstrass theorem there exists a subsequence $\{z'_n\}$ of $\{z_n\}$ converging to a point z_0 in D. Hence, by continuity of $P(z)$,

$$P(z_0) = P(\lim_{n\to\infty} z'_n) = \lim_{n\to\infty} P(z'_n) = \lim_{n\to\infty} 1/f(z'_n) = 0$$

contradicting our original assumption $P(z) \neq 0$ for all z. ⦰

THEOREM 5.19: The Mean Value Theorem of Gauss

Let $f(z)$ be analytic on and in the interior of a circle C centered at c. Then for the values $z(t)$ on C,

(5.19)
$$f(c) = \frac{1}{2\pi}\int_0^{2\pi} f(z(t))\,dt$$

Proof: Let C have the parametric form $z = c + re^{it}$ for $0 \leq t \leq 2\pi$. By Cauchy's formula **(5.11)**,

$$f(c) = \frac{1}{2\pi i}\int_C \frac{f(z)}{z-c}\,dz = \frac{1}{2\pi i}\int_0^{2\pi} \frac{f(z(t))}{re^{it}}(rie^{it})\,dt = \frac{1}{2\pi}\int_0^{2\pi} f(z(t))\,dt \quad ⦰$$

[†] Intuitively true. For the actual proof, write $P(z)$ in the form $z^n[a_n + Q(z)]$ where $Q(z) = a_{n-1}/z + a_{n-2}/z^2 + \cdots + a_0/z^n$, then show that $Q(z) \to 0$ as $|z| \to \infty$. Therefore $|P(z)| > \frac{1}{2}|a_n z^n|$ for $|z|$ large enough.

THEOREM 5.20: Maximum Modulus Theorem

If $f(z)$ is analytic on a closed region R consisting of a piecewise-smooth simple closed curve C and its interior, then the maximum of $|f(z)|$ on R exists and occurs at some point on the boundary.

Proof: The proof is in (4) steps.

(1) Suppose that $|f(z)|$ is unbounded on R. Then there exists a sequence $\{z_n\}$ in R such that $|f(z_n)| > n$ ($n = 1, 2, 3, \cdots$). Since R is closed and bounded, by the Bolzano-Weierstrass theorem (adapted to a closed and bounded region), the sequence $\{z_n\}$ has a subsequence $\{z_n'\}$ converging to a point z_0 in R. But since $f(z_n') \to \infty$, $f(z)$ fails to be continuous at z_0, a contradiction.

(2) Hence $|f(z)|$ is bounded, and if we take M to be its least upper bound, there exists c in R such that $|f(c)| = M$.

(3) Now if c lies on C, we are finished. Suppose c is an interior point of R. Some circle D with center c and positive radius lies in R, where $|f(z)| \leq M$ on D, as shown in Figure 5.13.

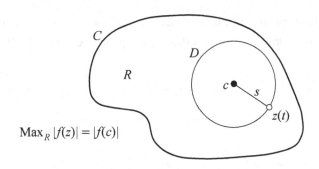

FIGURE 5.13

It will now be shown that $|f(z)| = M$ on D (parameterized by $z = c + se^{it}$, $0 \leq t \leq 2\pi$) by obtaining a contradiction of the mean value theorem of Gauss. Suppose that for some t_0 on $[0, 2\pi]$, $|f(z(t_0))| < M$. By continuity of $|f(z(t))|$ on D, there exists a neighborhood of reals $[t_1, t_2]$ of t_0 inside $[0, 2\pi]$ on which $|f(z(t))| < M$. Thus

$$\left| \int_0^{2\pi} f(z(t))\, dt \right| \leq \int_0^{2\pi} |f(z(t))|\, dt < \int_0^{2\pi} M\, dt = 2\pi M$$

Therefore,

$$|f(c)| = M > \left| \frac{1}{2\pi} \int_0^{2\pi} f(z(t))\, dt \right|$$

But this violates the Gauss mean value theorem, which requires that

$$f(c) = \frac{1}{2\pi} \int_0^{2\pi} f(z(t))\, dt$$

Thus, $|f(z)| = M$ for all z on circle D.

(4) Expand D to the boundary of R (since R is bounded). It follows that $|f(z)| = M$ for at least one point z on the boundary. ⬙

NOTE: The argument in step (3) can be used to show that $f(z) = M$ at all *interior points of R* since it is true on any circle D with center c. Thus, $f(z)$ is constant on the interior of D. Changing the location of c

in R, it can thus be shown that $f(z)$ is constant on R. Can this be true? See Problem 30. ⬩

By looking at the reciprocal function $g(z) = 1/f(z)$ for all z in R, one can use Theorem **5.20** to analyze the *minimum* of $|f(z)|$ on R. Provided $f(z) \neq 0$ in R, $g(z)$ is continuous, and by Theorem **5.20**, $g(z)$ assumes its maximum on the boundary, hence $f(z)$ assumes its minimum there. This proves:

THEOREM 5.21: Minimum Modulus Theorem

If $f(z)$ is analytic in a closed and bounded simply-connected region R with boundary C, and if $f(z) \neq 0$ on R, then the minimum of $|f(z)|$ exists and occurs at some point on the boundary.

THEOREM 5.22: (Morera)

If the integral of $f(z)$ is independent of path throughout an open, simply-connected region R, then f is analytic in R.

Proof: By Theorem **4.9** (proof considered in Problem 16 of Chapter 4), an anti-derivative of F of f was shown to exist throughout R. Thus, $F'(z) = f(z)$, which implies that $F''(z) = f'(z)$. ⬩

A concept using Cauchy's theorem and its consequences can be used to evaluate difficult real integrals (a method called *contour integration*). More on this will appear in Chapter 9. We give two examples of this procedure.

EXAMPLE 9

Show that for $a > 1$, $\displaystyle\int_0^{2\pi} \frac{a + \cos t}{a^2 + 2a\cos t + 1}\, dt = \frac{2\pi}{a}$.

SOLUTION

Let $C : z = a + e^{it}$, $0 \leq t \leq 2\pi$, be a unit circle centered at $z = a$. By Cauchy's integral formula **(5.11)**, with $f(z) = 1/z$,

$$\frac{2\pi i}{a} = \int_C \frac{1}{z} \cdot \frac{dz}{z-a} = \int_0^{2\pi} \frac{1}{a + e^{it}} \cdot \frac{1}{e^{it}} (i e^{it})\, dt$$

$$= i \int_0^{2\pi} \frac{1}{a + \cos t + i \sin t}\, dt$$

$$= i \int_0^{2\pi} \frac{a + \cos t - i \sin t}{(a + \cos t)^2 + \sin^2 t}\, dt$$

$$= \int_0^{2\pi} \frac{\sin t}{a^2 + 2a\cos t + 1}\, dt + i \int_0^{2\pi} \frac{a + \cos t}{a^2 + 2a\cos t + 1}\, dt$$

Equating imaginary parts on both sides, one obtains the desired result.

EXAMPLE 10

Show that for $a > b > 0$, $\displaystyle\int_0^{2\pi} \frac{dt}{a + b\cos t} = \frac{2\pi}{\sqrt{a^2 - b^2}}$.

SOLUTION

Since $z = e^{it}$, $dz = ie^{it}dt = izdt$. Also, recall the identity $\cos z = (e^{iz} + e^{-iz})/2$ for all z. Thus we make the following substitutions into the given integral I:

$$\cos t = \frac{e^{it} + e^{-it}}{2} = \frac{z + 1/z}{2} = \frac{z^2 + 1}{2z} \qquad \text{and} \qquad dt = \frac{dz}{iz}$$

Then

$$I = \int_C \frac{1}{a + b \cdot \dfrac{z^2 + 1}{2z}} \cdot \frac{dz}{iz} = \int_C \frac{2z}{2az + b(z^2 + 1)} \cdot \frac{dz}{iz} = \frac{1}{i} \int_C \frac{2dz}{bz^2 + 2az + b}$$

In order to factor the denominator, we use the quadratic formula for the equation $bz^2 + 2az + b = 0$:

$$z = \frac{-2a \pm \sqrt{4a^2 - 4b^2}}{2b} = \frac{-a \pm \sqrt{a^2 - b^2}}{b}$$

Thus, the two roots are $c_1 = \dfrac{\sqrt{a^2 - b^2} - a}{b}$ and $c_2 = -\dfrac{\sqrt{a^2 - b^2} + a}{b}$.
and it follows that

$$I = \frac{1}{i} \int_C \frac{2dz}{b(z - c_1)(z - c_2)}$$

Now c_1 lies inside C and c_2 lies outside (by testing the inequalities $c_1 < 1$ and $-c_2 > 1$), so we obtain by Cauchy's integral formula

$$I = \frac{1}{i} \int_C \frac{\dfrac{2}{b(z - c_2)}}{z - c_1} \, dz = \frac{1}{i} \cdot 2\pi i \cdot \frac{2}{b(c_1 - c_2)} = \frac{4\pi}{2\sqrt{a^2 - b^2}} = \frac{2\pi}{\sqrt{a^2 - b^2}}$$

as desired.

PROBLEMS

ALL CURVES OF INTEGRATION ARE TO HAVE THE POSITIVE ORIENTATION, UNLESS STATED OTHERWISE.

1. Evaluate the integral $\displaystyle\int_C \frac{z + 1}{z - 1} dz$ where C is the circle with center at $z = 0$ and radius

   (a) $\frac{1}{2}$

   (b) 2

2. Show that if C is the circle $|z| = 2$ then $\displaystyle\int_C \frac{e^{iz}}{z + 1} dz = 2\pi(\sin 1 + i\cos 1)$

3. Evaluate $\displaystyle\int_C \frac{e^z}{z + \frac{1}{2}} dz$ where C is the unit circle $|z| = 1$.

4. Evaluate $\displaystyle\int_C \frac{e^{z+1}}{z + \frac{1}{2}} dz$ where C is the unit circle $|z| = 1$.

5. Evaluate $\displaystyle\int_C \frac{\cos z}{z - \pi} dz$ when C is the:

   (a) unit circle $|z| = 1$,

   (b) circle $|z| = 4$.

6. Evaluate $\displaystyle\int_C \frac{\sin z}{z-\pi}\,dz$ where C is the unit circle $|z| = 1$.

7. Show that if C is the circle $|z| = 2$ the real part of the integral $\displaystyle\int_C \frac{e^{iz}dz}{z+1}$ equals $2\pi \sin 1$.

8. Show by direct calculation that if C is the circle $|z| = 2$, then $\displaystyle\int_0^{2\pi} \frac{z\,dz}{(z-1)^2} = 2\pi i$. [**Hint:** Use a circle D centered at $z = 1$ parameterized by $z = 1 + \frac{1}{2}e^{ti},\ 0 \le t \le 2\pi$.]

9. Use Cauchy's formula **(5.12)** on the integral in Problem 8 to arrive at the same result.

10. If C is the square having vertices $\pm i$ and $2 \pm i$, evaluate the integral $\displaystyle\int_C \frac{\cos \pi z}{(z^2 - 1)^2}$.

11. Evaluate $\displaystyle\int_C \frac{\sin z}{z^3 + 2z^2}\,dz$ if C is the unit circle $|z| = 1$.

12. Evaluate $\displaystyle\int_C \frac{z}{(z^2 + 4)^2}\,dz$ if C is the circle $|z + 2i| = 1$.

13. Evaluate $\displaystyle\int_C \frac{z + a}{(z^2 + 4)^2}\,dz$ if C is the circle $|z - 2i| = 1$.

14. Prove that if C is the circle $|z| = 2$, then for all real θ,

(a) $\displaystyle\frac{1}{2\pi i}\int_C \frac{e^{\theta z}}{z^2 - 1}\,dz = \sinh\theta$ (b) $\displaystyle\frac{1}{2\pi i}\int_C \frac{ze^{\theta z}}{z^2 - 1}\,dz = \cosh\theta$

15. Prove **(5.15)** by mathematical induction: It is already true for $n = 1$ by **(5.11)**; assume it is true for any $n > 1$, and by a technique similar to the proof of **(5.15)**, prove it for $n + 1$.

16. Show that if C is a closed curve enclosing the points c_1 and c_2, then

$$\int_C \frac{dz}{(z - c_1)(z - c_2)} = 0$$

17. Show that if C is a closed curve enclosing the points c_1, c_2, and c_3, then

$$\int_C \frac{dz}{(z - c_1)(z - c_2)(z - c_3)} = 0$$

In general, show that if C is a closed curve enclosing the points $c_1, c_2, c_3, \cdots, c_n$, then

$$\int_C \frac{dz}{(z - c_1)(z - c_2)(z - c_3)\cdots(z - c_n)} = 0$$

[**Hint:** For the general case, use partial fractions for the integrand (as in real variable methods for the same integral). Solve for A_1 in terms of z and $A_2, A_3, \cdots, A_n$, and take the limit of both sides as $z \to \infty$. Show that the integral equals $A_1 + A_2 + A_3 + \cdots + A_n$.]

18. Use **(5.19)** to evaluate $\displaystyle\int_0^{2\pi} \cos(\pi/3 + 2e^{ti})\,dt$.

19. Show that for any integer $n \ne 0$ $\displaystyle\int_C \frac{z^n}{(z - c)^n}\,dz = 2n\pi ci)$, where C is the circle $|z| = |c| + 1$.

20. Evaluate $\displaystyle\int_C \frac{z^6}{(z - \frac{1}{2})^5}\,dz$ where C is the unit circle $|z| = 1$.

21. Evaluate $\displaystyle\int_C \frac{\cos z}{(z - \pi)^6}\,dz$ where C is the unit circle $|z| = 1$.

22. Evaluate the integral $\int_0^{2\pi} \dfrac{dt}{a+b\sin t}$ in terms of a and b, where $a > b > 0$.

23. Evaluate the real integral $\int_0^{2\pi} \dfrac{a+\sin x}{a^2+2a\sin x+1}\,dx$ where $a > 1$ by using the procedure of Example 9.

24. Example for Green's Theorem (1). Let $P(x, y) = y/x$ and $Q(x, y) = 6x$. Further, consider the two paths C_1 and C_2, from $(0, 0)$ to $(2, 8)$, parameterized by t for $0 \le t \le 2$ and given by $x = t$, $y = 2t^2$ and $x = t$, $y = 4t$, respectively (see figure). Calculate:

(a) $\displaystyle\int_{C_1} (P\,dx + Q\,dy) \equiv \int_0^2 (P* \frac{dx}{dt} + Q* \frac{dy}{dt})\,dt$ where $P*(t) = P(t, 2t^2)$, $Q*(t) = Q(t, 2t^2)$

(b) $\displaystyle\int_{C_2} (P\,dx + Q\,dy)$

(c) $\displaystyle\int_C (P\,dx + Q\,dy)$ where C is the closed curve $C_1 - C_2$.

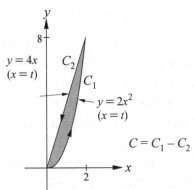

Figure for Problems 24, 25

25. Example for Green's Theorem (2). Calculate the following double integral over the region R enclosed by the curve C of Problem 24 (shaded in the figure):

$$\iint_R \left(\frac{\partial Q}{\partial x} - \frac{\partial P}{\partial y} \right) dx\,dy \equiv \int_0^2 \left[\int_{y=2x^2}^{y=4x} \left(\frac{\partial Q}{\partial x} - \frac{\partial P}{\partial y} \right) dy \right] dx$$

and compare your answer with that obtained in Problem 24(c).

26. Using Green's Theorem to Prove Cauchy's Theorem. Green's theorem may be stated as follows: For any two continuously differentiable functions $P(x, y)$ and $Q(x, y)$, and C any simple closed (piecewise-smooth) curve with interior R, then

$$\int_C (P\,dx + Q\,dy) = \iint_R \left(\frac{\partial Q}{\partial x} - \frac{\partial P}{\partial y} \right) dx\,dy$$

(This theorem is established in multivariable calculus for special cases.) From the definition of an integral over C (Chapter 4), the form of the left side of the equation in Green's theorem appears twice, given by

$$\int_C f(z)\,dz = \int_C (u\,dx - v\,dy) + i \int_C (v\,dx + u\,dy)$$

Prove that if $f(z)$ is analytic throughout C and its interior, then the above integral of $f(z)$ over C is zero.

NOTE: There is a gap in analysis between the type of regions used for Problems 24 and 25 and the most general type for which Green's theorem is usually applied, often glossed over in traditional calculus courses. In order to give a correct proof of Cauchy's theorem, this issue must be addressed.

27. Consider Liouville's theorem and the function $f(z) = \sin z$. This function is bounded for real numbers. Discuss why it cannot be bounded for complex numbers, using both a theoretical approach and the definition of $\sin z$ itself.

28. Find the maximum modulus of the analytic function $f(z) = |z - 2|$ over the neighborhood $|z| \leq 1$ and where it occurs. [**Hint:** Use geometry.]

29. Find the maximum modulus of $f(z) = \sin z$ over the neighborhood $|z| \leq 1$ and where it occurs.

30. Referring to the note on page 77 this is clearly not true for arbitrary functions $f(z)$. Examine the logic of the argument in step (3).

31. Prove Cauchy's integral formula if C is the boundary of a multiply-connected region R of order m (assume that $f(z)$ is analytic throughout R and on its boundary, and that c is an interior point of R).

32. Least Upper Bound Theorem for Subsets of Real Numbers.

Suppose S is any nonempty subset of real numbers, and that S is bounded above (there exists b such that for $x \in S$, $x \leq b$). Then there exists a **least upper bound** (a bound c such that no upper bound of S is less than c).

Complete the details of the following proof of this theorem:

(1) Let x be any point in S, and b an upper bound of S. For each integer $n > 1$, divide the interval $[x, b]$ into n subintervals of equal length. Since there are only finitely many of these subintervals, choose the greatest one (in terms of its endpoints) that contains points of S, and let x_n be any such point.

(2) If $\lambda = b - x$ (the length of $[x, b]$), then x_n has the property that $x_n + \lambda/n$ is an upper bound of S for each n.

(3) There exists a subsequence of $\{x_n\}$ that converges to a point c on $[x, b]$.

(4) Point c is an upper bound of S as a limit of upper bounds.

(5) Point c is the least upper bound of S as a limit of a sequence of points in S.

33. Greatest Lower Bound. Use the result of Problem 32 for the least upper bound to prove that if S is bounded below, then S has a greatest lower bound.

34. Heine-Borel Theorem for the Real Line. The statement of this theorem is as follows:

If a closed interval $[a, b]$ is covered by a family $\{U_\alpha : \alpha \in A\}$ of open sets of real numbers, then there exists a finite sub-family of these sets covering $[a, b]$.

(**NOTE:** A set U of real numbers is *open* iff for each point x in U, U contains an open interval whose center is x.)

Complete the details of the following proof of this theorem:

(1) The point $x = a$ is covered by some set U_λ for some λ in A, and U_λ contains an open interval (a, u), $u > a$. Thus, $[a, u)$ is covered by a finite number of sets from $\{U_\alpha : \alpha \in A\}$.

(2) Let $d \in [a, b]$ be the least upper bound of the set of all u such that $[a, u]$ is covered by finitely many sets from $\{U_\alpha : \alpha \in A\}$.

(3) Suppose that $d < b$. But d belongs to a set U_μ from $\{U_\alpha : \alpha \in A\}$. This is a contradiction, hence $d = b$, and the theorem is proved.

35. Heine-Borel Theorem for the Euclidean Plane. If a rectangle and its interior R is covered by a family of open sets $\{U_\alpha : \alpha \in A\}$ in the complex plane, then there exist a finite sub-family of these sets covering R. Prove this theorem, then extend it to any subset of the plane that is closed and bounded. (The topological term for a set that is closed and bounded is **compact**.)

36. Nested Sequences. Let $S_1 \supseteq S_2 \supseteq S_3 \supseteq \cdots \supseteq S_n \supseteq \cdots$ be a nested sequence of rectangles and their interiors in the complex plane. By considering a point $z_n \in S_n$ for each $n = 1, 2, \cdots$, show that any limit point z_0 of $\{x_n\}$ (a point to which a subsequence converges) belongs to all the rectangles in the nest. If the area of S_k converges to zero, show that z_0 is unique (no other point can lie in all the rectangles).

37. Give a counterexample to the result in Problem 36 if each S_n (n a positive integer) is the interior of a rectangle.

38. Carry out the required integrations to show that

$$\lim_{y \to 0} \int_0^1 \frac{2xy}{(x^2 + y^2)^2} \, dx \neq \int_0^1 \lim_{y \to 0} \frac{2xy}{(x^2 + y^2)^2} \, dx.$$

39. Counterexample in Real Variables. Define for all real numbers the function

$$f(x) = \begin{cases} x^2 \sin(1/x), & \text{if } x \neq 0 \\ 0, & \text{if } x = 0 \end{cases}$$

Show that the derivative $f'(x)$ exists for all real x, but that $f'(x)$ is discontinuous at $x = 0$.

40. Complete the details of the following proof of **(5.1)**.

(1) Let $z = z(t)$ for $a \leq t \leq b$ and $z_0 = z(t_0)$. Then if $g(t)$ is defined as $(dx/dt)^2 + (dy/dt)^2$,

$$l(z_0, z) = \int_{t_0}^t \sqrt{(dx/dt)^2 + (dy/dt)^2} \, dt = \int_{t_0}^t \sqrt{g(t)} \, dt$$

By a result in calculus, there exists u on $[t_0, t]$ such that $\int_{t_0}^t \sqrt{g(t)} \, dt = (t - t_0)\sqrt{g(u)}$

(2) Since $z(t) = x(t) + iy(t)$,

$$|z_0 - z| = \sqrt{[x(t) - x(t_0)]^2 + [y(t) - y(t_0)]^2}$$

(3) $\dfrac{l(z_0, z)}{|z_0 - z|} = \sqrt{\dfrac{(t - t_0)^2 g(u)}{[x(t) - x(t_0)]^2 + [y(t) - y(t_0)]^2}} = \sqrt{\dfrac{g(u)}{\left[\dfrac{x(t) - x(t_0)}{t - t_0}\right]^2 + \left[\dfrac{y(t) - y(t_0)}{t - t_0}\right]^2}}$

(4) $\lim_{t \to t_0} \dfrac{l(z_0, z)}{|z_0 - z|} = \lim_{t \to t_0} \sqrt{\dfrac{g(u)}{\left[\dfrac{x(t) - x(t_0)}{t - t_0}\right]^2 + \left[\dfrac{y(t) - y(t_0)}{t - t_0}\right]^2}} = \sqrt{\dfrac{g(t_0)}{[x'(t_0)]^2 + [y'(t_0)]^2}} = 1$

NOTE: Observe that the above argument requires $g(t) \neq 0$ for all t on $[a, b]$. Such curves are called **regular** in curve theory. However, this technicality has already been taken care of by our original definition of smooth curves, where it is required that $z'(t) \neq 0$ (and thus $g(t) \neq 0$). The remaining task is to prove **(5.1)** for piecewise smooth curves when z_0 is a "corner point", which will be left to the reader. ⬊

41. Proof of (5.9). Complete the details of the following proof ($f(z)$ is assumed to be analytic on C):

(1) $f(c + h) - f(c) = f'(c)h + \delta h$ where $\delta \to 0$ as $h \to 0$.

(2) Let $\varepsilon > 0$ be given. There exists $\lambda > 0$ such that if $|h| < \lambda$, $|f'(c)h + \delta h| < \varepsilon/l(C)$.

(3) By **(4.14)** $\left| \int_C f(c + h)dz - \int_C f(c)dz \right| \leq \int_C |f'(c)h + \delta h| |dz| \leq \dfrac{\varepsilon}{l(C)} \cdot l(C) = \varepsilon$ for $|h| < \lambda$.

(4) $\lim \int_C f(c + h)dz = \int_C f(c)dz$ as $h \to 0$.

42. Proof of (5.10). Complete the details of the following proof ($f(z)$ is assumed to be analytic, C is bounded away from $z = 0$ (that is, there exists $r > 0$ such that for all $z \in C$, $|z| \geq r$), and $g(z) = z^n$).

(1) For convenience, let $G_h(z) = [g(z + h) - g(z)]/h - g'(z)$. The result that must be proved then becomes

$$\lim_{h \to 0} \int_C f(z)G_h(z)dz = 0$$

(2) If $n > 0$ then $G_h(z) = \dfrac{(z+h)^n - z^n}{h} - nz^{n-1} = \dfrac{z^n + nz^{n-1}h + h^2 P(z,h) - z^n}{h} - nz^{n-1} = hP(z,h)$ where

$P(z, h)$ is a polynomial of degree $n - 2$ in z and h (for $n \geq 2$). Then since $P(z, h)$ is bounded if z and h are, we can let P be a bound for $|P(z, h)|$ for z on C and $|h| < 1$, and M one for $|f(z)|$ on C. Then by **(4.14)**,

$$\left| \int_C f(z) G_h(z) dz \right| = \left| \int_C f(z) h P(z,h) dz \right| \leq \int_C |f(z)| |hP(z,h)| |dz| \leq |h| MPl(C)$$

The result for this case then follows.

(3) The proof for $n < 0$ is similar, but involves more algebra. Let n be replaced by $-n$. Then

$$G_h(z) = \frac{1}{h} \left[\frac{1}{(z+h)^n} - \frac{1}{z^n} \right] - \frac{-n}{z^{n+1}} = \frac{1}{h(z+h)^n} - \frac{1}{hz^n} + \frac{n}{z^{n+1}}$$

$$= \frac{z^{n+1} - z(z+h)^n + nh(z+h)^n}{h(z+h)^n z^{n+1}} = \frac{z^{n+1} + (nh - z)(z+h)^n}{h(z+h)^n z^{n+1}}$$

Thus, $G_h(z) = \dfrac{hQ(z,h)}{(z+h)^n z^{n+1}}$ where $Q(z, h)$ is a polynomial in z and h.

(4) By hypothesis, $|z| \geq r$ and by the triangle inequality, if $|h| < \frac{1}{2}r$, then $|z + h| \geq |z| - |h| \geq r - \frac{1}{2}r = \frac{1}{2}r$. Let Q and M be bounds for $|Q(z, h)|$ and $|f(z)|$ as before. Then by **(4.14)**,

$$\left| \int_C f(z) G_h(z) dz \right| = \left| \int_C f(z) \frac{hQ(z,h)}{(z+h)^n z^{n+1}} dz \right| \leq \int_C |f(z)| \frac{|h||Q(z,h)|}{|z+h|^n |z|^{n+1}} |dz| \leq |h| \frac{2^n MQl(C)}{r^{2n+1}}$$

which proves the result for $n < 0$.

43. Criticize the following argument proposed as a simpler proof of **(5.10)**, and which supposedly generalizes it to arbitrary $g(z)$ (such that $g(z)$ and $g'(z)$ are continuous).

(1) By the approximation theorem **(3.6)**, for any z on C, $g(z + h) = g(z) + g'(z)h + \delta h$
where $\delta \to 0$ as $h \to 0$; $\delta = \delta(z)$ is a function of z for each fixed $h \neq 0$.

(2) $\delta(z) = \dfrac{g(z+h) - g(z)}{h} - g'(z)$. It follows that $\delta(z)$ is continuous and that the integral $\int_C f(z)\delta(z)dz$

exists for each fixed h.

(3) Let $\varepsilon > 0$ and let M be an upper bound for $|f(z)|$ on C. There exists $\lambda > 0$ such that for $|h| < \lambda$, $|\delta(z)| < \dfrac{\varepsilon}{Ml(C)}$ for each all z on C. Therefore,

$$\left| \int_C f(z) \cdot \frac{g(z+h) - g(z)}{h} dz - \int_C f(z) g'(z) dz \right| \leq \int_C |f(z)| |\delta(z)| |dz| \leq M \cdot \frac{\varepsilon}{Ml(C)} \cdot l(C) = \varepsilon$$

which proves the result.

6

INFINITE SERIES:

TAYLOR'S THEOREM

An infinite series of complex numbers is formally the same as its counterpart for real numbers. However, because of the implications of Taylor's theorem in complex variables, the theory becomes significantly different, and much more powerful. Methods for summing difficult infinite series emerge as a result, among other applications, as we shall see.

SEQUENCES AND SERIES

Even though the complex variable version of sequences and series parrots the real variable version, it is helpful to go over the basic ideas. A **sequence** in $\mathbb{C}$ is a function that maps the natural numbers into $\mathbb{C}$ (the complex numbers). For convenience, we let z_n denote the complex number associated with the natural number n. An example is

$$1 + \tfrac{1}{2}i, \ \tfrac{1}{2} + \tfrac{2}{3}i, \ \tfrac{1}{3} + \tfrac{3}{4}i, \ \cdots, \ \tfrac{1}{n} + \tfrac{n}{n+1}i, \cdots$$

where $z_1 = 1 + \tfrac{1}{2}i, \ z_2 = \tfrac{1}{2} + \tfrac{2}{3}i, \ z_3 = \tfrac{1}{3} + \tfrac{3}{4}i,$ and so on.

A sequence $\{z_n\}$ is **convergent** iff its members eventually become "arbitrarily close" to some complex number c as $n \to \infty$, and $\{z_n\}$ is then said to **converge** to c. Mathematically, this means that for any real $\varepsilon > 0$ however small, there is a sufficiently large real number N such that for all $n > N$, $|z_n - c| < \varepsilon$. As with real sequences, this event is denoted $\lim_{n \to \infty} z_n = c$ (equivalently, $z_n \to c$ as $n \to \infty$). A sequence that does not converge is said to **diverge**.

It is routine to show that if z_n is broken up into its real and imaginary parts $x_n + iy_n$, then

(6.1) $$z_n + iy_n \to a + bi \quad \text{iff} \quad x_n \to a \quad \text{and} \quad y_n \to b$$

Note that in the example given above, $\lim_{n \to \infty} z_n = \lim_{n \to \infty} \left(\dfrac{1}{n} + i \cdot \dfrac{n}{n+1} \right) = 0 + i \cdot 1 = i$

Given a sequence $\{z_n\}$, a **series** in $\mathbb{C}$ is the special *sequence of sums*, called **partial sums**, defined by

$$s_1 = z_1, \quad s_2 = z_1 + z_2, \quad s_3 = z_1 + z_2 + z_3, \quad \cdots, \quad s_n = z_1 + z_2 + z_3 + \cdots + z_n$$

The **infinite series** corresponding to this sequence is the infinite sum

$$\sum_{n=1}^{\infty} z_n = z_1 + z_2 + z_3 + z_4 + \cdots + z_n + \cdots$$

which is said to **converge** iff the sequence of partial sums $\{s_n\}$ converges, and the limit is called the **sum** of the series. This is equivalent to writing

$$\sum_{n=1}^{\infty} z_n = S \quad \text{iff} \quad \lim_{n \to \infty} s_n = S$$

If the limit does not exist, the series is said to **diverge**.

EXAMPLE 1

Since in real numbers $1 + \frac{1}{2} + \left(\frac{1}{2}\right)^2 + \left(\frac{1}{2}\right)^3 + \cdots = 2$, then the following infinite sum converges:

$$\left(1 + \frac{1}{2}i\right) + \left(\frac{1}{2} + \frac{1}{2^2}i\right) + \left(\frac{1}{2^2} + \frac{1}{2^3}i\right) + \cdots + \left(\frac{1}{2^n} + \frac{1}{2^{n+1}}i\right) + \cdots$$

$$= \left(1 + \tfrac{1}{2} + \tfrac{1}{4} + \cdots + 1/2^n + \cdots\right) + \left(\tfrac{1}{2} + \tfrac{1}{4} + \tfrac{1}{8} + \cdots + 1/2^{n+1} + \cdots\right)i$$

$$= 2 + i$$

When first introduced to infinite series, many students entertain the notion that as long as the nth term z_n of a series becomes arbitrarily small (that is, converges to zero), then since the amount being added as n increases is smaller and smaller, the sum does not change very much from that point on. So surely the infinite series must have a limit. But recall the harmonic series

$$1 + \frac{1}{2} + \frac{1}{3} + \frac{1}{4} + \frac{1}{5} + \cdots + \frac{1}{n} + \cdots$$

which is shown in calculus to diverge, having an *infinite sum*.

As you might recall, convergent infinite series do not behave like ordinary finite sums. For example, a standard calculus result is

$$1 - \frac{1}{2} + \frac{1}{3} - \frac{1}{4} + \frac{1}{5} - \frac{1}{6} + - \cdots = \ln 2$$

but the sum of exactly the same terms, after rearranging, is

$$1 + \frac{1}{3} - \frac{1}{2} + \frac{1}{5} + \frac{1}{7} - \frac{1}{4} + \frac{1}{9} + \frac{1}{11} - \frac{1}{6} - \cdots = \tfrac{3}{2}\ln 2$$

We show this later after we have covered a few concepts involving series. In fact, it can be shown in real analysis that if an alternating series converges only conditionally (it converges as given, but diverges if all the minus signs are changed to plus signs) it can be made to converge to *any given real number* by merely rearranging its terms! Thus, facts about series which appear to be obvious often need a close look before valid conclusions can be drawn.

REAL AND IMAGINARY PARTS OF INFINITE SERIES

Suppose the complex numbers c_n have the form $a_n + ib_n$ for each n. Then **(6.1)** implies

(6.2)
$$\sum_{n=1}^{\infty} c_n \equiv \sum_{n=1}^{\infty}(a_n + ib_n) = \sum_{n=1}^{\infty} a_n + i\sum_{n=1}^{\infty} b_n$$

and the series $\sum c_n$ converges iff its real and imaginary parts converge separately. This is equivalent to

$$(a_1 + ib_1) + (a_2 + ib_2) + (a_3 + ib_3) + \cdots = (a_1 + a_2 + a_3 + \cdots) + (ib_1 + ib_2 + ib_3 + \cdots)$$

At first glance **(6.2)**, seems to violate the rule about rearranging the terms of an infinite series. However, **(6.2)** is a *finite* rearrangement, thus valid. To prove this, suppose that s_n, s'_n, and s''_n are the partial sums of the above three series, respectively. Then, since these sums are finite, $s_n = s'_n + s''_n$ for each n, hence by the sum theorem on limits (assuming the limits exist),

$$\sum_{n=1}^{\infty}(a_n + ib_n) = \lim_{n\to\infty} s_n = \lim_{n\to\infty}(s'_n + s''_n) = \lim_{n\to\infty} s'_n + i\lim_{n\to\infty} s''_n = \sum_{n=1}^{\infty} a_n + i\sum_{n=1}^{\infty} b_n$$

Conversely, it can be shown that if *either* of the series of real numbers $\sum a_n$ or $\sum b_n$ diverges, then $\sum c_n$ diverges in $\mathbb{C}$.

A typical application of **(6.2)** is to determine the convergence of the series

$$\sum_{n=1}^{\infty}\left(\frac{1}{n^2} + \frac{i}{n^3}\right)$$

Because both the real series $\sum 1/n^2$ and $\sum 1/n^3$ converge as p-series with $p > 1$, then $\sum(1/n^2 + i/n^3)$ converges. On the other hand,

$$\sum_{n=1}^{\infty}\left(\frac{1}{n^2} + \frac{i}{n}\right)$$

diverges because $\sum 1/n$ diverges.

A list of some well-known results for real infinite series is provided by Table 1 below. This list of series will be found quite useful as a reference for later work. All but the last two results are standard in calculus (see Problems 50, 51 for the last two).

$$\sum_{n=1}^{\infty}\frac{(-1)^{n+1}}{n} = 1 - \frac{1}{2} + \frac{1}{3} - \frac{1}{4} + \cdots \pm \frac{1}{n} + \cdots = \ln 2 \qquad \sum_{n=0}^{\infty} r^n = 1 + r + r^2 + r^3 \cdots + r^n + \cdots = \frac{1}{1-r} \text{ if } |r| < 1$$

$$\sum_{n=1}^{\infty}\frac{1}{n^2} = 1 + \frac{1}{4} + \frac{1}{9} + \frac{1}{16} + \cdots + \frac{1}{n^2} + \cdots = \frac{\pi^2}{6} \qquad \sum_{n=1}^{\infty}\frac{1}{n^p} \text{ converges if } p > 1, \text{ diverges if } p \leq 1$$

$$\sum_{n=0}^{\infty}\frac{(-1)^n}{2n+1} = 1 - \frac{1}{3} + \frac{1}{5} + \cdots \pm \frac{1}{2n+1} + \cdots = \frac{\pi}{4} \qquad \sum_{n=1}^{\infty}\frac{\sin nx}{n^p} \text{ converges for all real } x \; (p > 0)$$

$$\sum_{n=0}^{\infty}\frac{1}{n!} = 1 + \frac{1}{2} + \frac{1}{6} + \frac{1}{24} + \cdots + \frac{1}{n!} + \cdots = e \qquad \sum_{n=1}^{\infty}\frac{\cos nx}{n^p} \text{ converges for } x \neq 2k\pi \; (p > 0)$$

TABLE 1

GEOMETRIC REPRESENTATION OF INFINITE SERIES

Recall the addition concept for complex numbers; since it is equivalent to that for vectors in the xy-plane, we obtain a diagram like that shown in Figure 6.1, where the sum of the complex numbers c_1 and c_2 is the

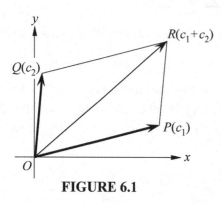

FIGURE 6.1

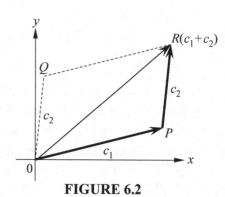

FIGURE 6.2

same as the sum of the vectors $OP(c_1)$ and $OQ(c_2)$, or OR, where OR is the diagonal of the parallelogram determined by OP and OQ. Figure 6.2 shows an equivalent diagram, where the vector OQ has been translated so that its starting point is at the endpoint of the vector $OP(c_1)$, and its endpoint is R (or $c_1 + c_2$).

This process can be extended to three complex numbers, c_1, c_2, and c_3, or to any finite number. One simply draws vectors that correspond to each individual complex number, positioned properly, as shown in Figure 6.3. The endpoint of the last vector represents the desired sum of the given numbers. The diagram exhibits a polygonal path, starting at 0 and ending at the final sum, where the vertices of the path are intermediate sums.

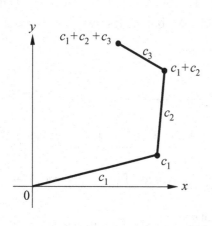

FIGURE 6.3

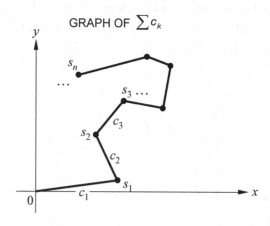

FIGURE 6.4

Thus the partial sums of a typical infinite series generate a polygonal path whose vertices are $s_1 = c_1$, $s_2 = c_1 + c_2$, $s_3 = c_1 + c_2 + c_3$, and so on. In this manner we obtain a graphical representation of an infinite series, called its **graph** (Figure 6.4). The convergence or divergence of a series thus takes on a geometric aspect, where as n increases, the endpoint s_n can be observed to either *diverge* (going off to infinity or randomly moving from point to point), as illustrated in Figure 6.5, or to *converge* (eventually settling down to a unique point), as illustrated in Figure 6.6.

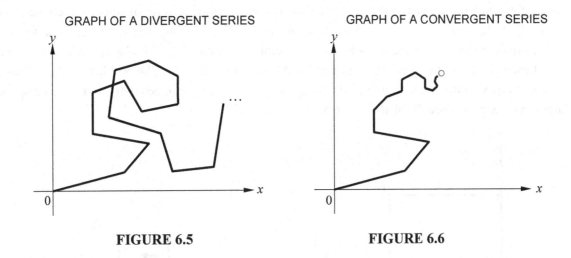

FIGURE 6.5   FIGURE 6.6

Certain properties of infinite series also have visual aspects in these diagrams. Take for example the theorem (proved below) that requires the terms of a convergent series to have zero limit. Thus, the lengths of the vectors c_n become arbitrarily small, as seen in Figure 6.6. If their lengths do not converge to zero and continue to remain greater than some fixed value, the divergence of the series appears in its graph, as seen in Figure 6.5. Consider a specific example of a divergent series,

$$\sum_{n=1}^{\infty} ni^n = i - 2 - 3i + 4 + 5i - 6 - 7i + - \cdots$$

whose graph is shown in Figure 6.7. It is represented by a spiral with its tail going off to infinity, telling us that the series diverges.

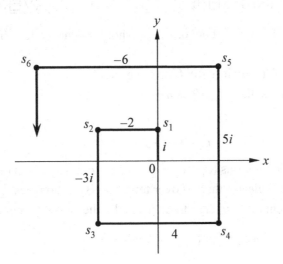

FIGURE 6.7

This method of representation is particularly effective in the case of a *geometric series*. For example, consider the two series

$$\sum_{n=0}^{\infty} i^n \quad \text{and} \quad \sum_{n=0}^{\infty} (\tfrac{2}{3}\, i)^n$$

The graph of the first one, which is divergent, consists of a *square* (Figure 6.8), where the partial sums are periodic, lying on a fixed square for all n. The second is represented by a spiral whose tail converges to the value $\tfrac{9}{13} + \tfrac{6}{13}\, i$ as shown in Figure 6.9. (Can you verify this? You need to use the formula for the complex valued geometric series; see below.) In general, the graph of $\sum c^n$ is a spiral, with congruent angles at each vertex (of measure $\pi \pm \mathrm{Arg}\, c$) and with sides of length $|c|^n$. The problems below include a few interesting examples, including one involving the golden rectangle in geometry and the corresponding *golden infinite series*. (See Problems 14–16.)

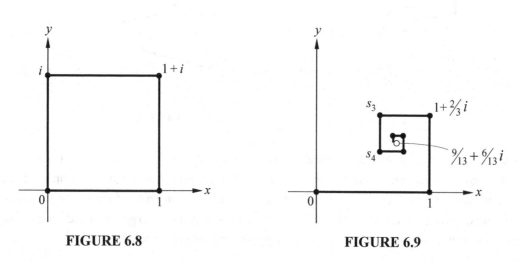

FIGURE 6.8   FIGURE 6.9

BASIC THEOREMS FOR INFINITE SERIES: TESTS FOR CONVERGENCE

The first result, apparently trivial but useful nevertheless, should seem familiar from calculus:

THEOREM 6.3 (Necessary Condition for Convergence)

If the infinite series $\sum c_n$ converges, then $c_n \to 0$ as $n \to \infty$.

Proof: For $n > 1$,

$$c_n = s_n - s_{n-1}$$

where s_n is the sum of the first n terms $(c_1 + c_2 + \cdots + c_{n-1} + c_n)$, and s_{n-1} is the sum of the first $n - 1$ terms. The difference is clearly the left over term, c_n. By hypothesis, $\sum c_n$ converges so we can suppose its partial sums s_n converge to S. Then, by the sum theorem on limits,

$$\lim c_n = \lim (s_{n-1} - s_n) = S - S = 0$$

as $n \to \infty$. ⧷

THEOREM 6.4 (Linear Combination Rule)

If $\sum c_n$ and $\sum d_n$ are convergent series, then for arbitrary numbers λ and μ, $\sum (\lambda c_n + \mu d_n)$ converges to the sum $\lambda \sum c_n + \mu \sum d_n$.

Proof: Let s_n, s'_n, and s''_n be the partial sums of $\sum(\lambda c_n + \mu d_n)$, $\sum c_n$, and $\sum d_n$, respectively. Since these sums are finite, $s_n = \lambda s'_n + \mu s''_n$. Then by the theorems on limits,

$$\sum(\lambda c_n + \mu d_n) = \lim_{n\to\infty} s_n = \lim_{n\to\infty}(\lambda s'_n + \mu s''_n) = \lambda \lim_{n\to\infty} s'_n + \mu \lim_{n\to\infty} s''_n = \lambda\sum c_n + \mu\sum d_n. \; \backslash\backslash$$

The next result establishes the theorem that if a series converges absolutely, then the series itself converges (as is true for real series but potentially harder to prove for complex series). We use the real variables concept that any bounded series of positive terms must converge [*Proof:* If $\sum a_n < M$, or $s_n < M$, then since s_n is monotonically increasing and bounded, $\lim s_n$ exists and equals the least upper bound of $\{s_n\}$]. In the following, let $c_n = a_n + ib_n$ for each n.

THEOREM 6.5 (Absolute Convergence)

If a series $\sum c_n$ converges absolutely, that is, if $\sum|c_n| \equiv \sum\sqrt{a_n^2 + b_n^2}$ converges, then $\sum c_n$ converges.

Proof: Suppose that $\sum\sqrt{a_n^2 + b_n^2} = S$. Consider the individual (real) series $\sum|a_n|$ and $\sum|b_n|$. The partial sums of these two real series of non-negative terms are bounded by S, and thus converge. Hence $\sum a_n$ and $\sum b_n$ converge absolutely as real series, so they converge. By **(6.2)**, $\sum c_n$ converges. $\backslash\backslash$

As in real analysis, the terms of an absolutely convergent series may be rearranged without affecting its sum, unlike the alternating harmonic series which is not absolutely convergent. For example, since the following p-series for $p > 1$ converges absolutely,

$$1 - \frac{1}{2^2} + \frac{1}{3^2} - \frac{1}{4^2} + \frac{1}{5^2} + - \cdots = 1 + \frac{1}{3^2} - \frac{1}{2^2} + \frac{1}{5^2} + \frac{1}{7^2} - \frac{1}{4^2} + \frac{1}{9^2} + \frac{1}{11^2} - \frac{1}{6^2} + - \cdots$$

EXAMPLE 2

Show that the series $\displaystyle\sum_{n=1}^{\infty} \frac{e^{ni}}{n^2}$ is absolutely convergent, therefore convergent.

SOLUTION

Since $e^{ni} = \cos n + i\sin n$ and $|e^{ni}| = (\cos^2 n + \sin^2 n)^{1/2} = 1$, then

$$\sum_{n=1}^{\infty}\left|\frac{e^{ni}}{n^2}\right| = \sum_{n=1}^{\infty}\frac{1}{n^2}$$

which converges to $\pi^2/6$ (Table 1). By Theorem 6.5, the given series converges.

EXAMPLE 3

Consider the two alternating series

(a) $\displaystyle\sum_{n=1}^{\infty}(-1)^{n+1}\left(\frac{1}{n} + \frac{i^n}{n}\right)$

(b) $\displaystyle\sum_{n=1}^{\infty}(-1)^{n+1}\left(\frac{1}{n} + \frac{(-1)^n i}{n}\right)$

Show that the one in (a) converges, and find its sum, and show that the one in (b) diverges.

SOLUTION

(a) Displaying the series termwise, one obtains for the nth partial sum

$$(1+i) - \left(\frac{1}{2} - \frac{1}{2}i\right) + \left(\frac{1}{3} - \frac{1}{3}i\right) - \left(\frac{1}{4} + \frac{1}{4}i\right) + \left(\frac{1}{5} + \frac{1}{5}i\right) - \left(\frac{1}{6} - \frac{1}{6}i\right) + \cdots + (-1)^{n+1}\left(\frac{1}{n} + \frac{i^n}{n}\right)$$

$$= (1+i) - \frac{1}{2}(1-i) + \frac{1}{3}(1-i) - \frac{1}{4}(1+i) + \frac{1}{5}(1+i) - \frac{1}{6}(1-i) + \cdots - (-1)^n \frac{1}{n}(1+i^n)$$

We notice that formally the terms begin to repeat after the fourth term, so we look at s_{4k} for $k = 1$, $2, \cdots (4k \le n)$. Thus

$$s_n = s_{4k} + \varepsilon_n = [1 - \frac{1}{2} + \frac{1}{3} - \frac{1}{4} + \frac{1}{5} - \cdots - \frac{1}{4k}]$$

$$+ [\frac{1}{2} - \frac{1}{4} + \frac{1}{6} - \frac{1}{8} + \cdots + \frac{1}{4k}] + i[1 - \frac{1}{3} + \frac{1}{5} - \frac{1}{7} + \cdots - \frac{1}{4k-1}] + \varepsilon_n$$

where ε_n denotes the sum of the remaining 1 to 3 terms if $4k < n$. Hence $\varepsilon_n \to 0$ and from Table 1

$$\sum_{n=1}^{\infty} (-1)^{n+1}\left(\frac{1}{n} + \frac{i^n}{n}\right) = \lim_{n\to\infty} s_n = \ln 2 + \tfrac{1}{2}\ln 2 + i \cdot \pi/4 = \ln 2\sqrt{2} + \pi i/4$$

(b) In partial sums, the coefficients of i in the series sum to

$$-1 - \frac{1}{2} - \frac{1}{3} - \frac{1}{4} - \frac{1}{5} - \cdots$$

which diverges. Therefore, the series diverges.

EXAMPLE 4

Show that if the terms of the alternating harmonic series

$$1 - \frac{1}{2} + \frac{1}{3} - \frac{1}{4} + \frac{1}{5} - \frac{1}{6} + - \cdots$$

(which converges to $\ln 2$) are rearranged to produce the series

$$1 + \frac{1}{3} - \frac{1}{2} + \frac{1}{5} + \frac{1}{7} - \frac{1}{4} + \frac{1}{9} + \frac{1}{11} - \frac{1}{6} + - \cdots$$

the new series converges to $\frac{3}{2}\ln 2$.

SOLUTION

A special method is used here to analyze the rearrangement: Take half of the partial sum s_n of the alternating harmonic sequence and write it in the form $0 + \frac{1}{2} + 0 - \frac{1}{4} + 0 + \frac{1}{6} + \cdots$, then place the resulting terms directly below the original partial sum:

$$1 - \frac{1}{2} + \frac{1}{3} - \frac{1}{4} + \frac{1}{5} - \frac{1}{6} + \frac{1}{7} - \frac{1}{8} + \frac{1}{9} - \frac{1}{10} + \frac{1}{11} - \frac{1}{12} + \cdots \quad = \ln 2 \ (s_n)$$

$$0 + \frac{1}{2} + 0 - \frac{1}{4} + 0 + \frac{1}{6} + 0 - \frac{1}{8} + 0 + \frac{1}{10} + 0 - \frac{1}{12} + \cdots \quad = \frac{1}{2}\ln 2 \ (\tfrac{1}{2}s_n)$$

$$+ \rule{10cm}{0.4pt}$$

$$1 + 0 + \frac{1}{3} - \frac{1}{2} + \frac{1}{5} + 0 + \frac{1}{7} - \frac{1}{4} + \frac{1}{9} + 0 + \frac{1}{11} - \frac{1}{6} + \cdots = \text{new series } (s'_n)$$

If we add the terms in each column, the result is the partial sum of the rearranged series. If the terms of the rearranged series are taken in groups of four, each group starting with zero, the partial sums of the rearranged series is the sum of these groups ± three or fewer terms (whose limit is zero). The rearranged partial sum thus converges to the sum of the two series above, or to $\ln 2 + \frac{1}{2}\ln 2 + 0 = \frac{3}{2}\ln 2$.

Recall the geometric series from calculus: If $|a| < 1$ then

$$\sum_{n=0}^{\infty} a^n = \frac{1}{1-a}$$

It would be interesting to show basically the same thing for a complex number c, with $|c| < 1$.

THEOREM 6.6 (Geometric Series)

For any complex number c having absolute value less than 1,

(6.6)
$$\sum_{n=0}^{\infty} c^n = 1 + c + c^2 + c^3 + c^4 + \cdots + c^n + \cdots = \frac{1}{1-c}$$

Proof: By algebra in $\mathbb{C}$, $(1-c)(1 + c + c^2 + c^3 + \cdots + c^n) = 1 - c^{n+1}$. If s_n is the nth partial sum of the original series, then

$$(1-c)s_n = 1 - c^{n+1} \qquad \text{or} \qquad s_n = \frac{1 - c^{n+1}}{1-c}$$

Since $|c|$ is a positive (or zero) real number < 1, $|c|^{n+1} \to 0$ as $n \to \infty$, and it follows that

$$\lim_{n \to \infty} s_n = \frac{1 - 0}{1-c} = \frac{1}{1-c} \quad \diagdown$$

There is also a *comparison test* in complex variables, like its counterpart in real variables. It states essentially that if each term of a given series is, in absolute value, less than or equal to the terms of a real convergent series, then the given series converges. For example, suppose we are given the series $\sum e^{nz}$, where z is any complex number with real part $x < 0$. Since $|e^{nz}| = |e^x|^n = (e^{-a})^n$ where $x = -a$, $a > 0$, then with $e^{-a} < 1$, we can conclude that because $\sum(e^{-a})^n$ converges as a geometric series, $\sum |e^{nz}|$ converges, and the given series converges absolutely, thus converges. In general:

THEOREM 6.7 (Comparison Test)

Suppose that for all sufficiently large n, $|c_n| \le |d_n|$, and suppose the series $\sum |d_n|$ converges. Then the series $\sum c_n$ converges absolutely, hence converges. If $|c_n| \ge |d_n|$ and the series $\sum |d_n|$ diverges, then $\sum c_n$ diverges absolutely (but may still converge).

Proof: This follows immediately from the comparison test for real series: Let $a_n = |c_n|$ and $b_n = |d_n|$ (real numbers). By hypothesis, $\sum b_n$ converges. Since

$$0 \le a_n \le b_n$$

then $\sum a_n$ converges (the comparison test in calculus). Thus $\sum c_n$ converges absolutely, and by Theorem 6.5, $\sum c_n$ converges. For the second part, suppose $\sum |d_n|$ diverges. Since the partial sums of a divergent series of positive terms is unbounded, and since $|c_n| \ge |d_n|$, those of $\sum |c_n|$ are unbounded and the latter series diverges (absolutely). $\diagdown$

THEOREM 6.8 (Ratio Test)

Suppose that $\lim_{n \to \infty} \left| \frac{c_{n+1}}{c_n} \right| = a$. Then if $a < 1$, the series $\sum c_n$ converges, and if $a > 1$, $\sum c_n$ diverges. If $a = 1$, no conclusion can be drawn.

Proof: The proof involves the three cases cited in the hypothesis.

(1) $a < 1$. Choose a real number b such that $0 \le a < b < 1$. Then with $a_n = |c_n|$, by definition of limit it follows that since a_{n+1}/a_n converges to a, which is less than b, then for all suffi-

ciently large $n > N$, $\dfrac{a_{n+1}}{a_n} < b$. Then it follows that

$$\frac{a_{n+1}}{a_n} \cdot \frac{a_{n+2}}{a_{n+1}} \cdot \frac{a_{n+3}}{a_{n+2}} \cdots \frac{a_{n+p}}{a_{n+p-1}} < b \cdot b \cdot b \cdots b \ \ (p \text{ factors}) \qquad \text{or} \qquad \frac{a_{n+p}}{a_n} < b^p$$

Thus, for any fixed $n > N$ and integer $p > 0$, $a_{n+p} < a_n b^p$ and

$$|c_{n+p}| < |c_p| b^p = d b^{n+p}$$

where $d = |c_p|/b^n$. So for m large enough,

$$|c_m| < d b^m \quad (m = n + p)$$

Since $b < 1$, the series $\sum d b^m$ converges as a constant times a geometric series. By the comparison test **(6.7)**, $\sum c_m$ converges.

(2) $a > 1$. We must have $a_n > 0$ for all sufficiently large n since the limit of a_{n+1}/a_n exists, and $a_{n+1}/a_n > 1$ for all n sufficiently large. Thus $a_{n+1} > a_n$, $a_{n+2} > a_{n+1} > a_n$, $a_{n+3} > a_{n+2} > a_n$, $\cdots$, $a_m > a_n$ for $m = n + p$ where a_n is a positive constant. Therefore, $\lim a_m \neq 0$ and $a_n = |c_n|$ does not converge to zero. By Theorem 6.3, $\sum c_n$ diverges.

(3) $a = 1$. The series can either converge, or diverge, as shown by the following two real series, both with $|c_{n+1}|/|c_n|$ (equal to either $n^2/(n^2 + 1)$ or to $n/(n + 1)$, which coverges to 1):

$$\sum_{n=1}^{\infty} \frac{1}{n^2} \qquad \text{and} \qquad \sum_{n=1}^{\infty} \frac{1}{n}$$

We mention two results that are useful for evaluating infinite series, involving manipulating series algebraically. The first one **(6.9)**, is valid in general, while the second **(6.10)**, requires the absolute convergence of each of the two the given series. We omit the proofs.

(6.9) $$\sum c_n + \sum d_n = \sum (c_n + d_n)$$

(6.10) Cauchy Product $$\sum c_n \cdot \sum d_n = c_1 d_1 + (c_1 d_2 + c_2 d_1) + (c_1 d_3 + c_2 d_2 + c_3 d_1) + \cdots$$

POWER SERIES FOR ANALYTIC FUNCTIONS

As in real analysis, the function e^z is the limit of a power series for each complex number z. As we will establish later,

$$e^z = 1 + z + \frac{z^2}{2!} + \frac{z^3}{3!} + \cdots + \frac{z^n}{n!} + \cdots$$

The above infinite series is an example of a **power series**, defined in general as any series of the form

$$a_0 + a_1(z - c) + a_2(z - c)^2 + a_3(z - c)^3 + \cdots + a_n(z - c)^n + \cdots$$

where z is a complex number.

The number c is called the **center**, and one speaks of the series as being **computed** or **expanded about** c. It will be shown momentarily that if a power series converges for any particular value of $z \neq c$, then it has an entire **region of convergence** of the form $|z - c| < r$ (the interior of a circle with radius r centered at c) plus, in some cases, points on the boundary as well. The maximal such r is called the **radius of convergence**. The theory of power series ranks among the most important concepts in real and complex analysis.

If a power series converges for all z in some neighborhood, then it defines a function $f(z)$ that is continuous and differentiable in that neighborhood (proof of this involves a concept known as *uniform convergence*, covered in the next chapter). One can start with a power series in order to define some function, or one can start with an analytic function and expand it into a power series, as we shall see in the next section. For example, it can be shown that the power series

$$z + (1 + \tfrac{1}{2})z^2 + (1 + \tfrac{1}{2} + \tfrac{1}{3})z^3 + (1 + \tfrac{1}{2} + \tfrac{1}{3} + \tfrac{1}{4})z^4 + \cdots$$

is convergent for all $|z| < 1$ (see Problem 29) and, accordingly, it defines an analytic function in that region. In this case, the function is not one of the familiar ones, like sine or cosine. You can see that this opens up a new method for constructing a large variety of analytic functions with very little effort (many of which are of little importance, however).

On the other hand, all the basic functions we introduced in Chapter 2 have power series expansions. For convenience, they are listed below (see Table 2). It is somewhat remarkable that the familiar series one encounters in calculus for the basic functions in real variables are identical to those for their counterparts in complex variables. These results are all applications of Taylor's theorem, which is taken up in the next section. It has the same basic characteristics as the theorem by the same name you learned in calculus, but there is a huge theoretical difference between the two.

$$\frac{1}{1+z} = 1 - z + z^2 - z^3 + z^4 + \cdots + (-1)^n z^n + \cdots \qquad (|z| < 1)$$

$$\sin z = z - \frac{z^3}{3!} + \frac{z^5}{5!} - \frac{z^7}{7!} + \cdots + \frac{(-1)^n z^{2n+1}}{(2n+1)!} + \cdots \qquad (\text{all } z)$$

$$\cos z = 1 - \frac{z^2}{2!} + \frac{z^4}{4!} - \frac{z^6}{6!} + \cdots + \frac{(-1)^n z^{2n}}{(2n)!} + \cdots \qquad (\text{all } z)$$

$$e^z = 1 + z + \frac{z^2}{2!} + \frac{z^3}{3!} + \frac{z^4}{4!} + \cdots + \frac{z^n}{n!} + \cdots \qquad (\text{all } z)$$

$$\sinh z = z + \frac{z^3}{3!} + \frac{z^5}{5!} + \frac{z^7}{7!} + \cdots + \frac{z^{2n+1}}{(2n+1)!} + \cdots \qquad (\text{all } z)$$

$$\cosh z = 1 + \frac{z^2}{2!} + \frac{z^4}{4!} + \frac{z^6}{6!} + \cdots + \frac{z^{2n}}{(2n)!} + \cdots \qquad (\text{all } z)$$

$$\ln(1+z) = z - \frac{z^2}{2} + \frac{z^3}{3} - \frac{z^4}{4} + \cdots + \frac{(-1)^{n+1} z^n}{n} + \cdots \qquad (|z| < 1)$$

TABLE 2

EXAMPLE 5

Use the ratio test (Theorem **6.8**) to show that the series for $\ln(1+z)$ in Table 2 converges for $|z| < 1$.

SOLUTION

Let $c_n = (-1)^{n+1} z^n / n$ (the nth term of the series). The ratio test requires taking the limit of $|c_{n+1}/c_n|$ as $n \to \infty$. Since $|c_n| = |z|^n / n$, we have

$$\left| \frac{c_{n+1}}{c_n} \right| = \frac{\dfrac{|z|^{n+1}}{n+1}}{\dfrac{|z|^n}{n}} = \frac{|z|^{n+1}}{n+1} \cdot \frac{n}{|z|^n} = \frac{n}{n+1} |z| \to |z| \qquad (\text{as } n \to \infty)$$

Thus $\lim |c_{n+1}| / |c_n| = a = |z|$. By the ratio test (**6.8**), the series converges if $|z| < 1$.

A power series with center c that converges in an open neighborhood N of c may be differentiated at all points of N, and the derivative equals the result of differentiating each term of the series, which is called *termwise differentiation*. Although this requires the series to be *uniformly convergent* in N, we choose to keep things as elementary as possible in this chapter, so this topic will be postponed. For now it will be assumed, as proved in Chapter 7, that if a series converges for $|z - c| < r$, then it may be differentiated termwise. For example, if

$$f(z) = a_0 + a_1(z-c) + a_2(z-c)^2 + a_3(z-c)^3 + \cdots + a_n(z-c)^n + \cdots$$

then

$$f'(z) = a_1 + 2a_2(z-c) + 3a_3(z-c)^2 + \cdots + n a_n(z-c)^{n-1} + \cdots$$

A similar concept is true for integration:

$$\int f(z)dz = a_0(z-c) + \frac{a_1(z-c)^2}{2} + \frac{a_2(z-c)^3}{3} + \cdots + \frac{a_n(z-c)^{n+1}}{n+1} + \cdots + C$$

These concepts are useful in applications of series, and appear often. As an example, we might note that the termwise differentiation of the series representing $\sin z$ in Table 2 results in the series representing $\cos z$, showing the consistency of these concepts.

A final theorem before we tackle Taylor series is very important; it enables us to quickly find the region of validity for power series (in which they may be differentiated or integrated termwise). This theorem generalizes the special case shown in Example 5.

THEOREM 6.11 (Radius of Convergence)

Let $\sum a_n(z-c)^n$ be a power series, and suppose that the **radius of convergence** is defined as

(6.11)
$$\lim_{n \to \infty} \left| \frac{a_n}{a_{n+1}} \right| = r$$

if the limit exists. Then if $r > 0$ the series converges (absolutely) for all z such that $|z - c| < r$, and diverges for $|z - c| > r$. On the boundary $|z - c| = r$, the series can either converge or diverge.

Proof: This follows directly from the ratio test. Consider the term $b_n \equiv a_n(z-c)^n$ for each n. By the ratio test we know that if $\lim |b_{n+1}/b_n| < 1$, the series converges, and if $\lim |b_{n+1}/b_n| > 1$ the series diverges. But since $r > 0$,

$$\lim_{n\to\infty} \left| \frac{b_{n+1}}{b_n} \right| = \lim_{n\to\infty} \left| \frac{a_{n+1}(z-c)^{n+1}}{a_n(z-c)^n} \right| = \lim_{n\to\infty} \left| \frac{a_{n+1}}{a_n} \right| |z-c| = \frac{|z-c|}{r}$$

Thus if $|z-c|/r < 1$ the series converges, and if $|z-c|/r > 1$ it diverges. (This analysis also proves that if $r = 0$ the series converges only for $z = c$, and if $r = \infty$, it converges for all z.) Finally for $|z-c| = r$, then $\lim |b_{n+1}/b_n| = 1$ and according to the ratio test, the series may either converge or diverge. ◥

EXAMPLE 6

Find the radius of convergence for the power series

$$\sum_{n=0}^{\infty} \frac{z^n}{3^n - 50}$$

and determine the *exact* region of convergence; that is, find precisely those values of z for which the series converges.

SOLUTION

The radius of convergence is given by

$$r = \lim_{n\to\infty} \left| \frac{\frac{1}{3^n - 50}}{\frac{1}{3^{n+1} - 50}} \right| = \lim_{n\to\infty} \frac{3^{n+1} - 50}{3^n - 50} = 3$$

Thus the radius of convergence is $r = 3$. By Theorem **6.11** the series converges for $|z| < 3$ and diverges for $|z| > 3$. For points on the boundary $|z| = 3$, or $z = 3e^{i\theta}$ ($0 \le \theta \le 2\pi$), merely substitute this into the series in order to examination convergence at these points:

$$\sum \frac{(3e^{i\theta})^n}{3^n - 50} = \sum \frac{3^n(\cos\theta + i\sin\theta)^n}{3^n - 50} = \sum \frac{\cos n\theta + i\sin n\theta}{1 - 50/3^n}$$

by DeMoivre's theorem. In this case it suffices to consider whether the nth term converges to zero. We have

$$\left| \frac{\cos n\theta + i\sin n\theta}{1 - 50/3^n} \right| = \frac{\sqrt{\cos^2 n\theta + \sin^2 n\theta}}{|1 - 50/3^n|} = \frac{1}{|1 - 50/3^n|} \to 1 \quad \text{as} \quad n \to \infty$$

We then conclude that the series diverges for all points on the circle $|z| = 3$, and the exact region of convergence is the interior of the circle (for $|z| < 3$). (See Figure 6.10.)

EXAMPLE 7

Find the exact region of convergence for the power series $\sum_{n=1}^{\infty} \frac{(z-i)^n}{n}$.

SOLUTION

Since $n/(n+1) \to 1$ as $n \to \infty$, the radius of convergence is 1, and the series converges for $|z - i| < 1$ and diverges for $|z - i| > 1$. For $|z - i| = 1$, let $z - i = e^{i\theta}$. By substitution, the series is

$$\sum \frac{e^{ni\theta}}{n} = \sum \frac{\cos n\theta + i\sin n\theta}{n} = \sum \frac{\cos n\theta}{n} + i\sum \frac{\sin n\theta}{n}$$

The two series $\sum (\cos n\theta)/n$ and $\sum (\sin n\theta)/n$ converge for $0 < |\theta| \leq \pi$ (see Table 1), leaving the value $\theta = 0$ $(z = 1 + i)$ to examine. But if $z = 1 + i$ the power series reduces to the divergent harmonic series. Thus the series converges for all z for which $|z - i| \leq 1$ and $z \neq 1 + i$, as shown in Figure 6.10.)

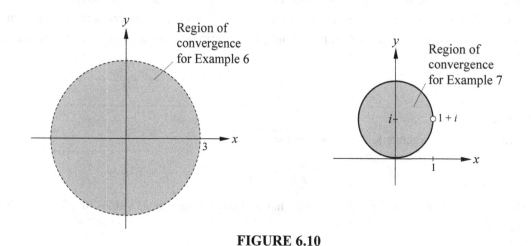

FIGURE 6.10

TAYLOR'S THEOREM FOR COMPLEX VARIABLES

Recall that Taylor's theorem in calculus states that if $f(x)$ has continuous derivatives of all orders on some neighborhood N of $x = a$ (that is, for $-r < x - a < r, r > 0$), then for each x in N there can be found a value c between x and a such that for any positive integer n,

$$f(x) = f(a) + \frac{f'(a)}{1!}(x - a) + \frac{f''(a)}{2!}(x - a)^2 + \frac{f^{(3)}(a)}{3!}(x - a)^3 + \cdots + \frac{f^{(n)}(a)}{n!}(x - a)^n + R_n(c)$$

where

$$R_n(c) = \frac{f^{n+1)}(c)(x - c)^n}{n!}(x - a)$$

(called a *remainder*). Then it was shown for certain functions $f(x)$ and for certain neighborhoods (intervals) of a that the remainder $R_n(c)$ has zero limit as $n \rightarrow \infty$, which then led to the *Taylor series expansion* of $f(x)$ about $x = a$:

$$f(x) = \sum_{n=0}^{\infty} \frac{f^{(n)}(a)(x - a)^n}{n!}$$

When this can be done, $f(x)$ is said to be *analytic* in that neighborhood.

Not all differentiable functions over the reals have a Taylor series expansion, even if the derivatives of all orders exist and are continuous (as required by Taylor's theorem). The standard example appears in Problem 32 in case you have not seen it. A remarkable result for complex variables is that *there is no remainder term for Taylor series*. The single assumption that $f(z)$ is differentiable in a neighborhood automatically guarantees a Taylor series expansion valid in that neighborhood, as we shall see.

Let's start with a neighborhood N centered at c and any point z in the interior of N, (z is a specific point (constant) for now. Also, let w lie on the boundary C of N, as indicated in Figure 6.11. If $f(z)$ is a complex function that is merely continuous on N and C, then the integral $\int f(w)\,dw$ exists along C. This is also true for the function $f(w)/(w-z)^n$, for each integer n, (since $w-z \neq 0$ for w on C). Note that if the radius of N is r and the distance from z to c is s, then as w varies on C, the quantity $|z-c|/|w-c| = s/r$ is constant, and has a value less than 1; this will be important for what follows.

With some algebraic manipulation, we obtain

$$\frac{1}{w-z} = \frac{1}{w-c-(z-c)} = \frac{1}{w-c} \cdot \frac{1}{1-\dfrac{z-c}{w-c}} = \frac{1}{w-c} \cdot \frac{1}{1-W}$$

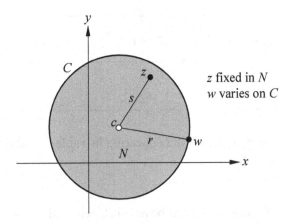

z fixed in N

w varies on C

FIGURE 6.11

where $W = (z-c)/(w-c)$, with $|W| = \dfrac{s}{r} < 1$. Now use the geometric series for $(1-W)^{-1}$:

$$\frac{1}{w-z} = \frac{1}{w-c}\left(1 + W + W^2 + W^3 + \cdots + W^n + \cdots\right)$$

This result can be written

(6.12)
$$\frac{1}{w-z} = \frac{1}{w-c} + \frac{z-c}{(w-c)^2} + \frac{(z-c)^2}{(w-c)^3} + \frac{(z-c)^3}{(w-c)^4} + \cdots$$

(This relation is to be noted since it will be used again in Chapter 8.) Now multiply both sides by $f(w)/2\pi i$ and integrate over C by termwise integration. This will give us the relation

$$\frac{1}{2\pi i}\int_C \frac{f(w)}{w-z}\,dw = \frac{1}{2\pi i}\int_C \frac{f(w)}{w-c} + \frac{1}{2\pi i}\int_C \frac{f(w)(z-c)}{(w-c)^2} + \frac{1}{2\pi i}\int_C \frac{f(w)(z-c)^2}{(w-c)^3} + \cdots$$

NOTE: Problem 36 shows how to derive this result without assuming the series can be integrated termwise. ✎

Now, since $(z-c)^n$ is constant for each n, it can be factored out. The above formula then takes on the following form:

(6.13)
$$\frac{1}{2\pi i}\int_C \frac{f(w)}{w-z} = \sum_{n=0}^{\infty} a_n(z-c)^n \qquad \text{where } a_n = \frac{1}{2\pi i}\int_C \frac{f(w)}{(w-c)^{n+1}}\,dw$$

This relation is valid for a function $f(w)$ that is merely continuous on C. Thus the integral of $f(z)/(w-z)$ is given in terms of the integrals a_n appearing in **(6.13)**.

Let's now assume that $f(z)$ is analytic over $N \cup C$. Using Cauchy's formula, we obtain

$$\frac{1}{2\pi i}\int_C \frac{f(w)}{w-z} = f(z) \qquad \text{and} \qquad a_n = \frac{1}{2\pi i}\int_C \frac{f(w)}{(w-c)^{n+1}}\,dw = \frac{f^{(n)}(c)}{n!}$$

for each n. By substitution of these results into **(6.13)**, we obtain:

THEOREM 6.14 (Taylor's Theorem)

Let $f(z)$ be analytic at $z = c$, thus differentiable in a neighborhood N of c. Then for all z in N,

$$\textbf{(6.14)} \quad f(z) = f(c) + f'(c)(z-c) + \frac{f''(c)}{2!}(z-c)^2 + \frac{f^{(3)}(c)}{3!}(z-c)^3 + \cdots + \frac{f^{(n)}(c)}{n!}(z-c)^n + \cdots$$

EXAMPLE 8

Obtain the Taylor series expansion of $\sin z$ appearing in Table 2, and find the region for which the resulting infinite series is valid.

SOLUTION

The neighborhood N in Theorem 6.14 can be any disk centered at $z = 0$. The series in **(6.14)** requires the calculation of the derivatives $(\sin z)^{(n)}$ for $n = 0, 1, 2, 3, \cdots$, and their evaluation at $z = 0$. Thus,

$$\sin z, \qquad\qquad \sin 0 = 0 \qquad\qquad (\sin z)^{(3)} = -\cos z, \qquad -\cos 0 = -1$$
$$(\sin z)' = \cos z, \qquad \cos 0 = 1 \qquad\qquad (\sin z)^{(4)} = \sin z, \qquad \sin 0 = 0$$
$$(\sin z)'' = -\sin z, \qquad -\sin 0 = 0 \qquad\qquad (\sin z)^{(5)} = \cos z, \qquad \cos 0 = 1 \cdots$$

Direct substitution into **(6.14)** (with $c = 0$) yields

$$\sin z = 0 + \frac{1}{1!}z + \frac{0}{2!}z^2 + \frac{-1}{3!}z^3 + \frac{0}{4!}z^4 + \frac{1}{5!}z^5 + \cdots = z - \frac{z^3}{3!} + \frac{z^5}{5!} + \cdots$$

The region of validity is seen to be the entire complex plane. To see this, we calculate the radius of convergence:

$$r = \lim_{n\to\infty}\frac{1/n!}{1/(n+1)!} = \lim_{n\to\infty}\frac{(n+1)!}{n!} = \lim_{n\to\infty}(n+1) = \infty$$

The series is therefore valid everywhere. (When $c = 0$ in **(6.14)** as in this case, the series is called a **Maclaurin series**.)

A few examples will illustrate methods that are typically used for the calculation of difficult real-valued series. Consider, for example, the two series

$$\sum \frac{\sin n}{n} \qquad \text{and} \qquad \sum \frac{\cos n}{n}$$

Intuitively, since about half the terms are negative and have an absolute value $\leq 1/n$, one might guess that they each resemble a convergent alternating series and that therefore each converges (the actual proof of this is found in Problem 51). But the question is, to what *value* do they converge?

EXAMPLE 9

Show that for $0 < \theta < 2\pi$,

(a) $\displaystyle\sum_{n=1}^{\infty} \frac{\sin n\theta}{n} = \frac{1}{2}\pi - \frac{1}{2}\theta$

(b) $\displaystyle\sum_{n=1}^{\infty} \frac{\cos n\theta}{n} = \ln\left(\frac{1}{2}\csc\frac{1}{2}\theta\right)$

SOLUTION

The two series are, respectively, the real and imaginary parts of

$$\sum_{n=1}^{\infty} \frac{(\cos n\theta + i\sin n\theta)}{n} = \sum_{n=1}^{\infty} \frac{(\cos\theta + i\sin\theta)^n}{n} = \sum_{n=1}^{\infty} \frac{(e^{i\theta})^n}{n}$$

Thus we need to evaluate the series $\displaystyle\sum_{n=1}^{\infty} \frac{z^n}{n}$ where $z = e^{i\theta}$. But from Table 2, this is the Taylor series for

$-\ln(1-z)$ for $-1 < |z| < 1$. If we let $0 < r < 1$ and use the definition of the logarithm,

$$\sum_{n=1}^{\infty} \frac{(re^{i\theta})^n}{n} = -\ln(1 - r\cos\theta - ir\sin\theta) = -\ln|1 - r\cos\theta - ir\sin\theta| - i\,\mathrm{Arg}(1 - r\cos\theta - ir\sin\theta)$$

Since the series converges for $r = 1$, we can take the limit of both sides as $r \to 1$. (Technically, this requires Abel's boundary theorem for complex variables, proved in Chapter 7.) Thus, the real part converges to

$$-\ln|1 - \cos\theta - i\sin\theta| = -\ln\sqrt{1 - 2\cos\theta + \cos^2\theta + \sin^2\theta} = -\frac{1}{2}\ln(2 - 2\cos\theta)$$

For the imaginary part, it converges to $\mathrm{Arg}(1 - \cos\theta - i\sin\theta)$. Since $1 - \cos\theta \geq 0$, $1 - \cos\theta - i\sin\theta$ lies in either the first or fourth quadrants, or on the y-axis, its value φ lies between $-\pi/2$ and $\pi/2$. Hence

$$\mathrm{Arg}(1 - \cos\theta - i\sin\theta) = \varphi = \tan^{-1}\left(\frac{\sin\theta}{\cos\theta - 1}\right)$$

Equating real and imaginary parts,

(a) $\displaystyle\sum_{n=1}^{\infty} \frac{\sin n\theta}{n} = \tan^{-1}\left(\frac{\sin\theta}{\cos\theta - 1}\right) = \tan^{-1}(\cot\frac{1}{2}\theta) = \tan^{-1}[\tan(\frac{1}{2}\pi - \frac{1}{2}\theta)] = \frac{1}{2}\pi - \frac{1}{2}\theta.$

(b) $\displaystyle\sum_{n=1}^{\infty} \frac{\cos n\theta}{n} = -\frac{1}{2}\ln 2(1 - \cos\theta) = -\frac{1}{2}\ln 4\left(\frac{1 - \cos\theta}{2}\right) = -\frac{1}{2}\ln[4\sin^2\frac{1}{2}\theta] = \ln\left(\frac{1}{2}\csc\frac{1}{2}\theta\right).$

EXAMPLE 10

Evaluate the (real) series $g(\theta) = \displaystyle\sum_{n=0}^{\infty} \frac{\cos n\theta}{n!}$, which is observed to converge absolutely for all θ.

SOLUTION

Consider the complex series

$$G(\theta) = \sum_{n=0}^{\infty} \frac{\cos n\theta}{n!} + i\sum_{n=0}^{\infty} \frac{\sin n\theta}{n!} = \sum_{n=0}^{\theta} \frac{(e^{i\theta})^n}{n!}$$

Thus we need to evaluate $\displaystyle\sum_{n=0}^{\infty} \frac{z^n}{n!}$ for $z = e^{i\theta}$. But from Table 2, this series is just e^z, valid for all z. Hence

$$g(\theta) = \mathscr{R}G(\theta) = \mathscr{R}e^z = \mathscr{R}\exp(\cos\theta + i\sin\theta) = \mathscr{R}\,e^{\cos\theta}\mathrm{cis}(\sin\theta) = e^{\cos\theta}\cos(\sin\theta)$$

EXAMPLE 11

Show that $\sin\theta + \dfrac{\sin 3\theta}{3} + \dfrac{\sin 5\theta}{5} + \cdots = \dfrac{\pi}{4}$ *regardless of the value for θ, $0 < \theta < \pi$.*

SOLUTION

This series is the imaginary part of

$$\sum_{n=0}^{\infty} \frac{\cos(2n+1)\theta + i\sin(2n+1)\theta}{2n+1} = \sum_{n=0}^{\infty} \frac{(\cos\theta + i\sin\theta)^{2n+1}}{2n+1} = \sum_{n=0}^{\infty} \frac{z^{2n+1}}{2n+1} \equiv g(z)$$

with $z = e^{i\theta}$. Differentiate termwise to obtain

$$g'(z) = \sum_{n=0}^{\infty} z^{2n} = 1 + z^2 + z^4 + z^6 + \cdots = \frac{1}{1-z^2} \quad (|z| < 1)$$

Thus

$$g(z) = \int \frac{1}{1-z^2}\, dz = \int \left(\frac{\tfrac{1}{2}}{1+z} + \frac{\tfrac{1}{2}}{1-z} \right) dz = \tfrac{1}{2}\ln(1+z) - \tfrac{1}{2}\ln(1-z) + C$$

($C = 0$ since $g(0) = 0$.) We obtain $g(z) = \tfrac{1}{2}\ln\dfrac{1+z}{1-z}$ and $g(e^{i\theta}) = \tfrac{1}{2}\ln\dfrac{1+\cos\theta + i\sin\theta}{1-\cos\theta - i\sin\theta}$. Thus

$\tfrac{1}{2}\ln\dfrac{i\sin\theta}{1-\cos\theta} = \tfrac{1}{2}\ln(iu) = \tfrac{1}{2}\ln|iu| + \tfrac{1}{2}i\,\mathrm{Arg}\,\{iu\}$ where $u > 0$ (real), and the imaginary part is $\pi/4$.

An important comment is needed at this point in order to take care of what seems to be a discrepancy. In Example 9, the formula $\sum \sin n\theta/n = \tfrac{1}{2}\pi - \tfrac{1}{2}\theta$ was obtained for $0 < \theta < 2\pi$. Thus if $\theta \to 0$ the series approaches $\tfrac{1}{2}\pi$ as limit. Yet the series *evaluated at* $\theta = 0$ is zero! This must mean that the formula we obtained is incorrect. Certainly, something seems to be wrong. Was it our unusual method for obtaining the results in Examples 9, 10, and 11? However, as we will learn in the next chapter, functions defined by series can sometimes be discontinuous, and concluding that $\lim \sum f(z) = f(c)$ as $z \to c$ is incorrect, in general. (The next example shows another series having the same property of discontinuity which is easier to work with.) Results in the next chapter will show however, that our above conclusions are all correct.

EXAMPLE 12

Use the method of the last several examples to evaluate the series $\sum z(1-z)^n$ at various points in its region of convergence ($|z - 1| < 1$ and $z = 0$).

SOLUTION

The formula for geometric series yields (for $|z - 1| < 1$):

$$\sum_{n=0}^{\infty} z(1-z)^n = z\sum_{n=0}^{\infty}(1-z)^n = z \cdot \frac{1}{1-(1-z)} = z \cdot \frac{1}{z} = 1$$

So the series has the constant value 1 for $|z - 1| < 1$). For example

$$\sum_{n=0}^{\infty} \tfrac{1}{2}(\tfrac{1}{2})^n = \tfrac{1}{2}(1 + \tfrac{1}{2} + \tfrac{1}{4} + \tfrac{1}{8} + \cdots) = \tfrac{1}{2}\cdot 2 = 1 \quad \text{and} \quad \sum_{n=0}^{\infty} \tfrac{1}{3}(\tfrac{2}{3})^n = \tfrac{1}{3}(1 + \tfrac{2}{3} + \tfrac{4}{9} + \tfrac{8}{27} + \cdots) = \tfrac{1}{3}\cdot 3 = 1$$

The series converges at $z = 0$, so by the method of the previous examples, the limit of the series as $z \to 0$ which has the value 1, should be the value of the series at $z = 0$, which is certainly false!

One final issue will be taken up in this section, which will answer a question in connection with the definitions we made for the standard functions of complex variables in Chapter 2. Although they were motivated by logical reasoning, there was a nagging question about correctness. These functions are all analytic in some region that includes the real axis, or parts thereof. So the following theorem applies.

THEOREM 6.15

Suppose that $g(z)$ is analytic on the real axis, and that it coincides with a real function $f(x)$ when $z = x$ (real). Then $g(z)$ is unique.

Proof: Suppose that h is another function that is analytic on the real axis and coincides with $f(x)$ for real z. Let $z = c$ (real). By Taylor's theorem there exist complex numbers a_n and b_n such that for $|z - c| < q$ for some $q > 0$,

$$g(z) = \sum_{n=0}^{\infty} a_n (z - c)^n \qquad \text{and} \qquad h(z) = \sum_{n=0}^{\infty} b_n (z - c)^n$$

Let x be a real number within q of c. Then by hypothesis,

$$\sum_{n=0}^{\infty} a_n (x - c)^n = \sum_{n=0}^{\infty} b_n (x - c)^n = f(x)$$

(which proves that $f(x)$ has a Taylor series representation and is analytic as a real function at $x = c$). Thus, for $|x - c| < q$,

$$a_0 + a_1(x - c) + a_2(x - c)^2 + \cdots = b_0 + b_1(x - c) + b_2(x - c)^2 + \cdots$$

Setting $x = c$ shows that $a_0 = b_0$. Thus, cancelling a_0 and b_0,

$$a_1(x - c) + a_2(x - c)^2 + \cdots = b_1(x - c) + b_2(x - c)^2 + \cdots$$

Dividing by $x - c$ and setting $x = c$ again, we obtain $a_1 = b_1$. Continuing this process, it follows that $a_n = b_n$ for all n, which proves that $g(z) = h(z)$ for $|z - c| < q$. Since this is true for neighborhoods of all real numbers, $h(z) = g(z)$ wherever $g(z)$ has a valid Taylor series expansion. ⟍

Theorem 6.15 shows that $g(z)$ is unique for points near the x-axis. In Chapter 9 a method known as *analytic continuation* will show that $g(z)$ is unique throughout its domain R (if R is simply-connected). This is known as the *uniqueness theorem* for analytic functions.

EXAMPLE 13

In Chapter 2 the complex function $\sin z$ was defined as

$$\sin(x + iy) = \sin x \cosh y + i \cos x \sinh y$$

and it follows that $\sin z = \sin x$ for $z = x$, real. In Chapter 3 it was found that $\sin z$ is differentiable, thus analytic. By the uniqueness theorem, this is the only way to define $\sin z$ if you want it to be differentiable and to agree with $\sin x$ for real x.

The proof of the following theorem follows the same pattern as that for Theorem 6.15, so it will be omitted.

THEOREM 6.16 (Uniqueness Theorem for Taylor Series)

Suppose the infinite series $\sum a_n (z - c)^n$ converges to an analytic function $f(z)$ for $|z - c| < r$. Then $f^{(n)}(c)/n! = a_n$ and the given series is the Taylor series for $f(z)$ expanded about $z = c$.

REMOVABLE POINTS OF SINGULARITY

In complex variables, a **singular point** of a function $f(z)$ is any point $z = c$ for which $f(z)$ is *not analytic*, that is, non-differentiable. Often such points can be spotted easily. For example, $(z - i)^{-1}$ is not continuous at $z = i$, so this would be a singular point. A point of singularity can sometimes be *removed*, however. This is a descriptive term that means that the original function can be defined (or redefined) so that it becomes analytic at the singular point. (There is a similar procedure in calculus.) An illustration will clarify this concept for complex variables, and it also points out how some series can be given a "closed form".

Consider the series $1 - \dfrac{z^2}{3!} + \dfrac{z^4}{5!} - \dfrac{z^6}{7!} + \cdots$. This series converges for all z, and can be differentiated termwise. It thus represents an analytic function $f(z)$. But what function? (We are seeking a definition in terms of elementary functions that do not involve infinite series.)

Suppose we multiply this series by z; then the Taylor series for $\sin z$ emerges, and $zf(z) = \sin z$. So it follows that $f(z) \equiv \dfrac{\sin z}{z}$, which has a singularity at $z = 0$. Suppose we *define* the function

$$F(z) = \begin{cases} (\sin z)/z & \text{if } z \neq 0 \\ 1 & \text{if } z = 0 \end{cases}$$

(The motivation for this is the fact that $\lim \sin z/z = 1$ as $z \to 0$.) It will follow that $F(z)$ is not only continuous at $z = 0$, but it is differentiable there as well. To explicitly justify differentiability at 0, observe that

$$F'(0) = \lim_{h \to 0} \frac{F(0 + h) - F(0)}{h} = \lim_{h \to 0} \frac{(1 - h^2/3! + h^4/5! + \cdots) - 1}{h} = \lim_{h \to 0} \frac{-h^2/3! + h^4/5! + \cdots}{h} = 0$$

Thus, the derivative of $F(z)$ exists at $z = 0$, and at all other points. It is therefore analytic at all points in the complex plane (such functions are called **entire functions**), and the (unique) Taylor series expansion of $F(z)$ about $z = 0$ is the series we started with. The singular point of $(\sin z)/z$ at $z = 0$ is thus called **removable**, and the above procedure shows how to do it.

In general, we make the following definition.

DEFINITION   Suppose $f(z)$ has a singularity at an interior point $z = c$ in a region R but is analytic at all other points in R. The singularity at c is termed **removable** if $\lim [f(z)] = d$ exists as $z \to c$, and if, after one defines $f(c) = d$, $f(z)$ is analytic in R.

We anticipate a theorem whose proof must wait until Chapter 8, which involves the concept of Laurent series.

THEOREM 6.17 (Removable Singularities Theorem)

Suppose $g(z)$ is analytic at all points in a region R except at the single interior point $z = c$. If the limit of $g(z)$ exists as $z \to c$, then the singular point $z = c$ is removable.

NOTE: Theorem 6.17 is not true in real variables. It is true that if $f(x)$ is differentiable on an interval $[a, b]$ except at $c \in [a, b]$ and the limit $\lim f(x) = L$ exists as $x \to c$, then defining $f(c) = L$ will make $f(z)$

continuous at c. But $f(x)$ need not be *differentiable* at c. The standard counterexample is $f(x) = x \sin 1/x$ for $x \neq 0$, which is differentiable for all $x \neq 0$. Here, $\lim f(x) = 0$ as $x \to 0$, so if $f(0)$ is defined as 0, f will be continuous at $x = 0$, but it is not differentiable at $x = 0$ (consider the limit of $[f(0 + h) - f(0)]/h$ as $h \to 0$). ◣

PROBLEMS

In Problems 1–8, find the limits if they exist, and if not, explain why not.

1. $\displaystyle \lim_{n \to \infty} \left(\frac{\ln n}{n} + n \sin \frac{1}{n} i \right)$

2. $\displaystyle \lim_{n \to \infty} \left(\frac{6n + n^2 i}{n+1} + \frac{(1 - n^2)i}{n+2} \right)$

3. $\displaystyle \lim_{n \to \infty} \left(\tfrac{1}{2} + \tfrac{2}{3} i \right)^n$

4. $\displaystyle \lim_{n \to \infty} \left(\tfrac{1}{2} + i \right)^n$

5. $\displaystyle \lim_{n \to \infty} e^{n\pi i/2}$

6. $\displaystyle \lim_{n \to \infty} e^{n\pi i}$

7. $\displaystyle \lim_{n \to \infty} (-1)^n \cos(n\pi + i/n)$

8. $\displaystyle \lim_{n \to \infty} (-1)^n \sin(n\pi + i/n)$

9. REVIEW The tail of a series is the infinite sum of its terms from some point on. A series converges iff its tails converge to zero. That is, if a series converges and we are given $\varepsilon > 0$, there exists K such that for all $k > K$, $\left| \displaystyle \sum_{n=k}^{\infty} a_n \right| < \varepsilon$, and conversely. (See in this connection Problem 49.) For example, the tail of the alternating harmonic series $1 - \frac{1}{2} + \frac{1}{3} - \frac{1}{4} + \cdots$. from the kth term on is the infinite series represented by

$$\frac{1}{k} - \frac{1}{k+1} + \frac{1}{k+2} - \frac{1}{k+3} + \cdots$$

(if k is odd). How far does one have to go in order for the tail to have absolute value less than 0.005? [**Hint:** Show that the above expression is less than $1/k$ in absolute value.]

10. Graph the series $\displaystyle \sum_{n=0}^{\infty} (\tfrac{9}{10} i)^n$. Does the graph reveal the convergence/divergence property of this series?

11. Graph the series $\displaystyle \sum_{n=0}^{\infty} (\tfrac{3}{5} + \tfrac{4}{5} i)^n$. Does the graph reveal the convergence/divergence property of this series?

12. Estimate the sum of the series $\displaystyle \sum_{n=0}^{\infty} \left(\frac{1+i}{2} \right)^n$ from its graph. What is the exact sum?

13. An infinite staircase starting at $z = 0$ is constructed as shown in the figure. Find by limits how high the staircase rises if the size of each succeeding step (width of step and rise to the next step) is
(a) $\frac{2}{3}$ that of the preceding step (also find the limiting endpoint reached by the staircase),
(b) $\frac{99}{100}$ that of the preceding step (also find the limiting endpoint reached by the staircase).

14. The Golden Rectangle (prelude to Problem 16). The famous rectangle which the Greek geometers believed had the most appealing shape among all rectangles (which they called the **golden rectangle**) has sides in the ratio $\tau : 1$, where $\tau \ [\approx 1.618034]$ is the positive solution of the quadratic equation $x^2 - x - 1 = 0$

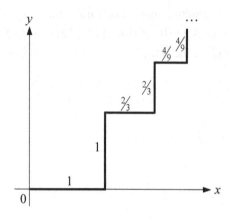

Figure for Problem 13

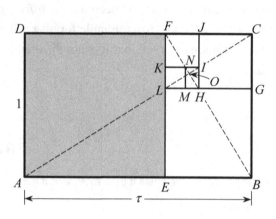

Figure for Problem 14

and, accordingly, has the properties $\tau^2 = \tau + 1$ and $\tau^{-1} = \tau - 1$. Thus, any power of τ can be expressed in the form $a\tau + b$. [Examples: $\tau^3 = \tau \cdot \tau^2 = \tau^2 + \tau = (\tau + 1) + \tau = 2\tau + 1$ and $\tau^{-2} = \tau^{-1}(\tau - 1) = 1 - \tau^{-1} = 1 - (\tau - 1)$ $= -\tau + 2$.] The number τ is called the **golden ratio**. The rectangle $ABCD$ shown in the figure for this problem, where $AB = \tau$ and $AD = 1$, is an example of a golden rectangle.

(a) Show that if a square $AEFD$ and its interior is cut off from rectangle $ABCD$, the result is another golden rectangle $EBCF$. (Obtain the relation $1/(\tau - 1) = \tau$.)

(b) What happens if we continue this process of cutting off squares from each successive golden rectangle? (The figure shows the cut-off points E, G, J, K, M, $\cdots$.)

(c) Let O be the intersection of diagonals AC and FB. Prove that O lies in the interiors of all the golden rectangles left after cutting off squares in the manner described in (b). [**Hint:** Prove that L and I lie on AC and that N and H lie on FB using similar triangles. Then show that O lies on segments LC and FH, in the interior of rectangle $LGCF$. The rest is by induction, since this same result is true for all the golden rectangles which follow.]

15. The Golden Spiral (prelude to Problem 16). Referring to problem 14, where O is the intersection of diagonals AC and FB, let A be the origin of a coordinate system whose x-axis is line AB and whose y-axis is line AD (see figure below).

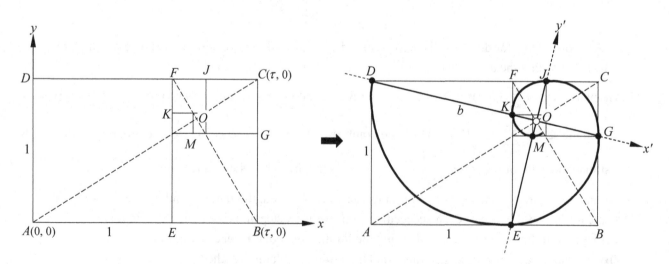

(a) Show that O is the complex number $\dfrac{2\tau + 1}{\tau + 2} + \dfrac{\tau + 1}{\tau + 2} i$.

[**Hint:** The coordinate equation of AC is $y = \tau/x$; find that for FB, then solve the system for x and y.]

(b) With O as origin and lines DO and EO as coordinate axes, and with $DO = b$, find the polar coordinates (r, θ) of D ($r > 0$) and the cut-off points $E, G, J, K,$ and M mentioned in Problem 14. [**Hint:** The first two points are $D(b, -\pi)$ and $E(b/\tau, -\pi/2)$.]

(c) Show that the points defined in (b) satisfy the polar equation $r = ce^{a\theta}$, where $c = b/\tau^2$ and $a = -(2/\pi)\ln\tau$. (This curve belongs to the class of **logarithmic spirals**, all having the distinctive property that for each point P on the curve, the polar radius OP makes a constant angle with the tangent at P.)

16. The Golden Infinite Series. The vertices of the graph of the geometric series

$$\sum_{n=0}^{\infty} \left(\frac{i}{\tau}\right)^n$$

lie on a golden logarithmic spiral centered at the point where the series converges. The easiest way to show this is to make a transformation: multiply the series by the factor $c = 1 - i$, and add i. The result is the series

$$i + \sum_{n=0}^{\infty} c\left(\frac{i}{\tau}\right)^n = i + c\cdot\frac{i}{\tau} + c\cdot\left(\frac{i}{\tau}\right)^2 + c\cdot\left(\frac{i}{\tau}\right)^3 + \cdots$$

The graph of this series is the result of rotating the original graph through an angle of $-45°$ (which is the argument of $1 - i$), multiplying by $\sqrt{2}$ (which is the length of $1 - i$), and translating the result up one unit (adding i to each point). See figure for illustration.

(a) Show that the vertices of the graph of the new series coincide with D and the cut-off points $E, G, J, \cdots$ mentioned in Problem 15.

(b) To what point does this new series converge? Prove your answer. [Compare with Problem 15(a).]

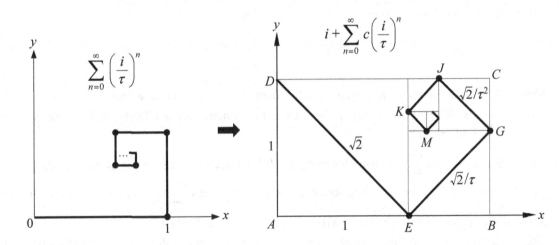

17. Show that $\displaystyle\sum_{n=1}^{\infty} \frac{i^n}{n^2}$ converges absolutely.

18. Show that $\displaystyle\sum_{n=1}^{\infty} \frac{i^n}{n}$ converges, but does not converge absolutely.

19. Show that $\displaystyle\sum_{n=1}^{\infty} \frac{i^n}{(1+i)^n}$ converges by the ratio test.

20. Show that $\displaystyle\sum_{n=1}^{\infty} \frac{(1+i)^n}{n^2}$ diverges by the ratio test.

21. Show that $\displaystyle\sum_{n=1}^{\infty} \frac{\sin nz}{n}$ converges for all z for which $\mathcal{I}(z) = 0$, but diverges elsewhere. (Use Table 1.)

22. Show that $\displaystyle\sum_{n=1}^{\infty} \frac{(3/5 + 4/5\, i)^n}{n}$ converges, but does not converge absolutely.

23. To what value does the series in Problem 22 converge? [**Hint:** See Example 9.]

24. Does the series $\displaystyle\sum_{n=1}^{\infty} \frac{(1+i)^{2n}}{2^n}$ converge or diverge?

25. Does the series $\displaystyle\sum_{n=1}^{\infty} \frac{(1+i)^{2n}}{3^n}$ converge or diverge?

26. Find the radius of convergence of the power series $\displaystyle\sum_{n=1}^{\infty} \left(\frac{1+i}{2}\right)^n \frac{z^n}{n^2}$.

27. Find the radius of convergence of the power series $\displaystyle\sum_{n=0}^{\infty} \frac{z^n}{3^{n+1}}$.

28. To what function $f(z)$ does the series of Problem 27 converge in its domain of convergence? Find those values on the boundary of the domain of convergence for which this series diverges. Are these values the same as the values of discontinuity for $f(z)$?

29. Show that the following series has a radius of convergence $r = 1$ and therefore converges to an analytic function:

$$z + (1 + \tfrac{1}{2})z^2 + (1 + \tfrac{1}{2} + \tfrac{1}{3})z^3 + \cdots + (1 + \tfrac{1}{2} + \tfrac{1}{3} + \cdots + \tfrac{1}{n})z^n + \cdots$$

30. Find the exact region of convergence for $\displaystyle\sum_{n=1}^{\infty} \frac{(z+2)^n}{(n+1)^3 4^n}$. (Be sure to examine points on the circle of convergence.)

31. Show that for $0 < |z| < 3$, $\displaystyle\sum_{n=1}^{\infty} \frac{z^n}{3^{n+2}} = \frac{z}{27 - 9z}$

32. The following continuous function is defined for all real x:

$$\begin{aligned} f(x) &= e^{-1/x^2} \quad \text{for } x \neq 0 \\ &= 0 \qquad\quad \text{for } x = 0 \end{aligned}$$

Show that this function has derivatives of all orders at every point (show this specifically at $z = 0$ and that the derivatives all equal zero at this point). Does this function have a Taylor series expansion at $z = 0$ that is valid in a neighborhood?

33. Unlike the real series $\displaystyle\sum_{n=1}^{\infty} \frac{x^n}{\sqrt{n}}$ which diverges at all but one boundary point of the interval $(-1, 1)$ of convergence, show that the corresponding series in complex numbers $\displaystyle\sum_{n=1}^{\infty} \frac{z^n}{\sqrt{n}}$ converges at all but one boundary point of the region of convergence. [**Hint:** Let $z = re^{i\theta}$. See Table 1.]

34. Use Taylor's theorem to find a MacLaurin series expansion for $\cosh z$. For what values of z is this valid?

35. Show that for $0 < \theta < \pi$, $\cos\theta + \tfrac{1}{3}\cos 3\theta + \tfrac{1}{5}\cos 5\theta + \cdots = \ln\sqrt{\cot\tfrac{1}{2}\theta}$.

36. With a_n defined as in (**6.14**), show that $f(z) = a_0 + a_1(z - c) + a_2(z - c)^2 + \cdots + a_n(z - c)^n + U_n$ where

$$U_n = \frac{1}{2\pi i} \int_C \frac{f(w)W^{n+1}}{w - z} dw$$

and that $U_n \to 0$ as $n \to \infty$. [**Hint:** Using ordinary algebra, $(w - z)^{-1} = (w - c)^{-1}[1 + W + W^2 + \cdots + W^n + W^{n+1}/(1 - W)]$; multiply by $f(w)/2\pi i$ and integrate, as in the method used for (**6.14**), and show that if B is a bound for $|f(w)(w - z)^{-1}|$ on C, (**4.14**) and (**4.15**) establish the result $|U_n| \leq Bs(s/r)^n \to 0$ as $n \to \infty$.]

37. As in real analysis, define the *binomial coefficients*

$$\binom{r}{n} = \frac{r(r-1)(r-2)(r-3)\cdots(r-n+1)}{n!}$$

where r is any real number. If r is a positive integer, the binomial theorem is the following rule for expanding the product $(x + y)^r$:

$$(x+y)^r = x^r + rx^{r-1}y + \binom{r}{2}x^{r-2}y^2 + \binom{r}{3}x^{r-3}y^3 + \cdots + \binom{r}{r}y^r \equiv \sum_{n=0}^{r}\binom{r}{n}x^{r-n}y^n$$

Using Taylor's theorem, show that for $|z| < 1$, $(1+z)^r = \sum_{n=0}^{\infty}\binom{r}{n}z^n$ (called the **binomial series**).

38. Calculate the first few binomial coefficients for $r = \frac{1}{2}$ and use the result of Problem 37 to show that for $|z| < 1$,

$$\sqrt{1+z} = 1 + \frac{z}{2} - \frac{1}{4}\frac{z^2}{2} + \frac{1\cdot3}{4\cdot6}\frac{z^3}{2} - \frac{1\cdot3\cdot5}{4\cdot6\cdot8}\frac{z^4}{2} + \frac{1\cdot3\cdot5\cdot7}{4\cdot6\cdot8\cdot10}\frac{z^5}{2} - \cdots - \frac{1\cdot3\cdot5\cdot(2n-3)}{4\cdot6\cdot8\cdot2n}\frac{(-z)^n}{2} + \cdots$$

39. Calculate the first few binomial coefficients for $r = -\frac{1}{2}$ and use the result of Problem 37 to show that for $|z| < 1$,

$$(1-z^2)^{-\frac{1}{2}} = 1 + \frac{1}{2}z^2 + \frac{1\cdot3}{2\cdot4}z^4 + \frac{1\cdot3\cdot5}{2\cdot4\cdot6}z^6 + \cdots + \frac{1\cdot2\cdot5\cdots(2n-1)}{2\cdot4\cdot6\cdots2n}z^{2n} \cdots$$

(a) By integrating both sides of the above equation, show that

$$\sin^{-1}z = \sum_{n=1}^{\infty}\frac{1\cdot3\cdot5\cdots(2n-1)}{2\cdot4\cdot6\cdots2n}\cdot\frac{z^{2n+1}}{2n+1} = z + \frac{1}{2}\frac{z^3}{3} + \frac{1\cdot3}{2\cdot4}\cdot\frac{z^5}{5} + \frac{1\cdot3\cdot5}{2\cdot4\cdot6}\cdot\frac{z^7}{7} + \cdots$$

(b) Find the function in closed form to which the following series converges (for $|z| < 1$)

$$\sum_{n=1}^{\infty}\frac{1\cdot3\cdot5\cdots(2n-1)}{2\cdot4\cdot6\cdots2n}z^n = \frac{1}{2}z + \frac{1\cdot3}{2\cdot4}z^2 + \frac{1\cdot3\cdot5}{2\cdot4\cdot6}z^3 + \cdots$$

(c) Show that

$$\frac{1}{4} + \frac{1}{2}\cdot\frac{3}{16} + \frac{1}{2}\cdot\frac{3}{4}\cdot\frac{5}{48} + \frac{1}{2}\cdot\frac{3}{4}\cdot\frac{5}{6}\cdot\frac{7}{128} + \frac{1}{2}\cdot\frac{3}{4}\cdot\frac{5}{6}\cdot\frac{7}{8}\cdot\frac{9}{320} + \cdots = \sqrt{2} - 1$$

40. After dividing by z and using termwise integration for infinite series, find the function in closed form to which the following series converges:

$$\sum_{n=1}^{\infty}nz^n \equiv g(z) \qquad (|z| < 1)$$

41. Using the result in Problem 40, show that

$$\frac{1}{3} - \frac{1}{9} + \frac{5}{3^5} - \frac{7}{3^7} + \frac{9}{3^9}\cdots + \frac{(-1)^n(2n+1)}{3^{2n+1}} + \cdots = \frac{3}{10}$$

[**Hint:** Let $z = i/3$.]

42. Find the function represented by the series $\sum_{n=0}^{\infty}(2n+1)z^{2n}$ as in Problem 40 and show that

$$\frac{3}{2} - \frac{7}{8} + \frac{11}{32} - \frac{15}{128} + \cdots + \frac{(-1)^{n+1}(4n-1)}{2^{2n+1}} + \cdots = \frac{22}{25}$$

[**Hint:** $z^2 = (\frac{1}{2} + \frac{1}{2}i)^2 = \frac{1}{2}i$.]

43. Suppose the MacLauren series for an analytic function $f(z)$ is $a_0 + a_1z + a_2z^2 + a_3z^3 + a_4z^4 + \cdots$. Show that if $f(z)$ is even [that is, $f(-z) = f(z)$ for all z in its domain] then $a_1 = a_3 = a_5 = \cdots = 0$ and the series consists of terms with even powers only. Similarly, show that if $f(z)$ is odd, the series consists of terms with odd powers only.

44. Show that $f(z) = \dfrac{\cos z - 1}{z^2}$ has a removable singularity at $z = 0$ and is analytic elsewhere. (You must show differentiablity.)

45. Show that $f(z) = \sin z / z^2$ has an essential singularity at $z = 0$, but is analytic elsewhere.

46. Show that $f(z) = z/\ln(z+1)$ has a removable singularity at $z = 0$. (You must show differentiability.)

47. Examine the hypothesis of Taylor's theorem and its proof in order to either prove, or disprove the following conjecture:

Suppose $f(z)$ is analytic in a neighborhood $N(z_0, r)$ for some $r > 0$ and has a Taylor series representation about z_0 as center that is valid in $N(z_0, r)$. Then if z_1 is the nearest point to z_0 for which $f(z)$ fails to be analytic, the Taylor series representation for $f(z)$ is valid in $N(z_0, s)$, where $s = |z_1 - z_0|$.

Problems 48–53 involve advanced concepts in real analysis, which have applications for complex variables.

48. **Cauchy Sequences.** A topic often given only a brief treatment in calculus is the concept of a *Cauchy sequence* and the role it plays in infinite series. A **Cauchy sequence** is any sequence of numbers $\{x_n\}$ (real or complex) such that for each $\varepsilon \geq 0$ there can be found a number K such that if $n > K$, $|x_n - x_{n+p}| < \varepsilon$ for all $p > 0$. An important theorem in real analysis states that *every Cauchy sequence converges*. You are to study the steps of the proof of this theorem for real numbers which appears below (it uses the Bolzano-Weierstrass theorem, Problem 31, Chapter 3), and to generalize it for complex numbers showing, that a *Cauchy sequence of complex numbers converges*.

(1) A Cauchy sequence is bounded: There exists K such that for $n > K$, $|x_n - x_{n+p}| < 1$ for all $p > 0$. If B is the maximum value of the numbers $x_1, x_2, x_3, \cdots, x_{K+1}$, then with $n = K + 1$,

$$|x_{n+p}| = |-x_{n+p}| = |x_n - x_{n+p} - x_n| \leq |x_n - x_{n+p}| + |-x_n| < 1 + B$$

for all $p > 0$. Thus, $|x_m| < B + 1$ for $m = n + p > n$, as well as for $m \leq n = K + 1$.

(2) By the Bolzano-Weierstrass theorem, the sequence $\{x_n\}$ has a subsequence $\{y_m\}$ that converges to a real number a. Let $\varepsilon > 0$ be arbitrarily small. Thus we can find a K such that for $m > K$, $|y_m - a| < \varepsilon/2$, and since $\{x_n\}$ is Cauchy, $|x_n - x_{n+p}| < \varepsilon/2$ for $n > K$ and $p > 0$. For each $m > K$, choose p such that $y_m = x_{n+p}$. Hence $\lim x_n = a$ since for all $n > K$,

$$|x_n - a| = |x_n - x_{n+p} + x_{n+p} - a| \leq |x_n - x_{n+p}| + |y_m - a| < \varepsilon/2 + \varepsilon/2 = \varepsilon$$

49. **Cauchy Sequences and Infinite Series.** The previous result (Problem 48) can be used to show that if all we know about an infinite series is that its "tails" are arbitrarily small (going out far enough in the series), then it can be concluded that the series converges. To be more specific, suppose that for any $\varepsilon > 0$ there exists K such that if $n > K$ then $|a_n + a_{n+1} + a_{n+2} + \cdots + a_{n+p}| < \varepsilon$ for all $p \geq 0$ (equivalent to requiring

$$|a_n + a_{n+1} + a_{n+2} + \cdots| \leq \varepsilon).$$

(a) Show that for the series $\sum a_n$ o f real numbers, the sequence of partial sums s_n constitute a Cauchy sequence, then $\sum a_n$ converges to some number a.

(b) Extend this result to include series of complex numbers.

50. **Abel's Theorem on Convergent Series for Real Variables.**

Suppose that an infinite series in real variables is of the form $\sum a_n b_n$ and that the partial sums of $\sum a_n$ are bounded. If, in addition, $\{b_n\} \to 0$ as $n \to \infty$ and is monotone non-increasing (that is, $b_1 \geq b_2 \geq b_3 \geq \cdots \geq b_n \geq b_{n+1} \geq \cdots$), the series converges.

Fill in the details and justify the following steps in the proof of this theorem. (It closely resembles the proof of Abel's boundary theorem to be presented in the next chapter.)

(1) To prove that the series converges to some real number a, we need only show that the tails of the series are, in absolute value, less than a given $\varepsilon > 0$ if we go out far enough (see Problem 49).

(2) Let $\varepsilon > 0$ and suppose that $|a_1 + a_2 + a_3 + \cdots + a_n| < M$ for all n.

(3) Using the notation $s_n = \sum_{k=1}^{n} a_k$, it follows that $a_m = s_m - s_{m-1}$ for each integer $m > 1$.

(4) For all $n >$ some K, $0 \leq b_n < \varepsilon/2M$.

(5) $\left| \sum_{k=n}^{n+p} a_k b_k \right| = \left| a_n b_n + a_{n+1} b_{n+1} + a_{n+2} b_{n+2} + a_{n+3} b_{n+3} + \cdots + a_{n+p} b_{n+p} \right|$

$$= |(s_n - s_{n-1})b_n + (s_{n+1} - s_n)b_{n+1} + (s_{n+2} - s_{n+1})b_{n+2} + (s_{n+3} - s_{n+2})b_{n+3} + \cdots + (s_{n+p} - s_{n+p-1})b_{n+p}|$$

$$\leq |s_{n-1}||b_n| + |s_n||b_n - b_{n+1}| + |s_{n+1}||b_{n+1} - b_{n+2}| + |s_{n+2}||b_{n+2} - b_{n+3}| + \cdots$$
$$+ |s_{n+p-1}||b_{n+p-1} - b_{n+p}| + |s_{n+p}||b_{n+p}|$$

$$< M[b_n + (b_n - b_{n+1}) + (b_{n+1} - b_{n+2}) + (b_{n+2} - b_{n+3}) + \cdots + (b_{n+p-1} - b_{n+p}) + b_{n+p}]$$

$$= M[2b_n] < \varepsilon$$

and thus the series $\sum_{n=1}^{\infty} a_n b_n$ converges.

51. Convergence of $\sum \sin nx/n^p$ and $\sum \cos nx/n^p$. Abel's theorem of Problem 50 can be used to show the convergence of these two series, as was assumed previously. Thus, letting $a_n = \sin nx$ and $b_n = 1/n^p$ for $p > 0$ (and similarly for $\cos nx/n^p$), it suffices to show that the partial sums of $\sum \sin nx$ and $\sum \cos nx$ are bounded. (Note that x must be restricted for the series involving $\cos nx/n^p$ since this series is unbounded if $x = 2k\pi$ where k is an integer and $p \leq 1$.) Let

$$S_k = \sum_{n=0}^{k} \sin nx \qquad \text{and} \qquad C_k = \sum_{n=0}^{k} \cos nx$$

(a) To show S_k is bounded for all x, let

$$O = \sin x + \sin 3x + \sin 5x + \sin 7x + \cdots + \sin(2k+1)x$$

$$E = \sin 2x + \sin 4x + \sin 6x + \sin 8x + \cdots + \sin(2kx)$$

Thus we can write

$$O = \tfrac{1}{2}[(\sin x + \sin 3x) + (\sin 3x + \sin 5x) + (\sin 5x + \sin 7x) + \cdots$$
$$+ (\sin(2k-1)x + \sin(2k+1)x)] + \tfrac{1}{2}\sin x + \tfrac{1}{2}\sin(2k+1)x$$

Using the sum-to-product identity for $\sin A + \sin B$ [$= 2\sin \tfrac{1}{2}(A+B)\cos \tfrac{1}{2}(A-B)$], we obtain

$$O = \sin \tfrac{1}{2}(4x)\cos \tfrac{1}{2}(2x) + \sin \tfrac{1}{2}(8x)\cos \tfrac{1}{2}(2x) + \sin \tfrac{1}{2}(12x)\cos \tfrac{1}{2}(2x) + \cdots$$
$$+ \sin \tfrac{1}{2}(4kx)\cos \tfrac{1}{2}(2x) + \sin \tfrac{1}{2}(2k+2)x \cos \tfrac{1}{2}(2kx)$$

$$= E\cos x + \sin(k+1)x \cos kx$$

In a similar manner, show that using the definition of E above, $E = O\cos x + \sin(k+1)x\cos(k-1)x - \sin(2k+1)x\cos x$.

(b) Obtain from these results the formula

$$(1 - \cos x)(O + E) = \sin(k+1)x\cos kx + \sin(k+1)x\cos(k-1)x - \sin(2k+1)x\cos x.$$

Thus $S_k = O + E$ is bounded (if $x = 2m\pi$ then $O + E = 0$).

(c) In the same manner as (a) and (b), show that C_k is bounded if $x \neq 2m\pi$ for $m = 0, \pm 1, \pm 2, \cdots$. (If $x = 2m\pi$, then $O + E = 2k + 2$ is unbounded).

(d) Show that each of the series $\sum \sin nx/n^p$ and $\sum \cos nx/n^p$ $(p > 0)$ converges for all real x (the latter provided $x \neq 2k\pi$).

52. Establish the absolute convergence of the last two series listed in Table 1 for $p \geq 2$ using the standard results for p-series in real variables.

53. A generalization of the alternating harmonic series $1 - \tfrac{1}{2} + \tfrac{1}{3} - \tfrac{1}{4} + \cdots$ can be obtained by considering a larger class of series, namely those represented by $\sum \varepsilon_n/n$, where $\varepsilon_n = \pm 1$, and the $\pm$ signs are determined according to some rule. A specific example would be $1 + \frac{1}{2} - \frac{1}{3} - \frac{1}{4} + \frac{1}{5} + \frac{1}{6} - \frac{1}{7} - \frac{1}{8} + \frac{1}{9} + \frac{1}{10} - \frac{1}{11} - \cdots$, which converges to $\pi/4 + \ln\sqrt{2}$. Unless the plus and minus signs are more or less randomly distributed, the series can diverge. Show that if for some constant C the number of plus signs in the first n terms does not exceed $n/2 + C$ for all sufficiently large n, then the series converges. [**Hint:** Use Abel's theorem.]

7

UNIFORM CONVERGENCE

The concept of uniform convergence involves sequences of *functions* rather than sequences of *numbers*. For example, we might consider the limit of z^n as $n \to \infty$, where z varies over some region. Here the sequence would be defined by the function $f_n(z) = z^n$ for all positive integers n. In general, $\lim f_n(z)$ depends on z; it may exist for certain values of z and not for others. If the limit does exist, it defines a function $f(z) = \lim f_n(z)$, called the *pointwise limit*. We pursue the properties of such limits in this chapter.

UNIFORM CONVERGENCE FOR SEQUENCES

In any limit involving a sequence of functions $\{f_n(z)\}$ with a pointwise limit $f(z)$, some natural questions arise. Suppose, for example, that each $f_n(z)$ is a continuous function on R for each n and that $\lim_{n \to \infty} f_n(z) = f(z)$ for each $z \in R$. One could ask whether $f(z)$ is also continuous on R. The same question concerns other properties, such as differentiability and integrability. The key to answering these questions is *uniform convergence*.

Two examples will show different outcomes for $\lim f_n(z)$, where, in both cases, $f_n(z)$ is continuous for each n on the same domain R.

EXAMPLE 1

Let $f_n(z) = z^n$ and consider the domain R consisting of the open disk $|z| < 1$ together with the single point $z = 1$. Find the limit function $f(z)$ on R.

SOLUTION

If $|z| < 1$ and $|z| = a$ where $0 < a < 1$, then $\lim |z| = \lim a^n = 0$. Thus $\lim z^n = 0$ if $|z| < 1$. But if $z = 1$, $\lim z^n = \lim 1^n = 1$. Here, $f(z)$ is defined on R and consists of the discontinuous function

$$f(z) = \begin{cases} 0 & \text{if } |z| < 1 \\ 1 & \text{if } z = 1 \end{cases}$$

EXAMPLE 2

Let $f_n(z) = z/n$ for $z \in R$, where R is the same region as in Example 1. Find the limit function $f(z)$.

SOLUTION

Since $|z|/n \leq 1/n$ for $|z| \leq 1$, $\lim |z|/n = 0$ and $\lim z/n = 0$ on R. Thus $f(z) = 0$ for all $z \in R$, a continuous function.

What distinguishes the different results in Examples 1 and 2? Certainly, one could say that the limit $z^n \to 0$ depends on z, while $z/n \to 0$ does not. To be more specific, the value $|f_n(z) - f(z)|$ approaches zero as $n \to \infty$ for each specific $z \in R$ in both cases. Thus for each $\varepsilon > 0$,

$$|f_n(z) - f(z)| < \varepsilon$$

for all n large enough, where $z \in R$. But in the case of $f_n(z) = z^n$, if z is allowed to vary in R the above inequality is violated; simply let $z = \sqrt[n]{\frac{1}{2}}$ for each $n \geq 2$ (which lies in R) and take $\varepsilon = \frac{1}{4}$. Since $f(z)$ is equal to either 0 or 1 for all $z \in R$, $|z^n - f(z)| = \frac{1}{2} \not< \varepsilon$. On the other hand, in Example 2 where $f(z) = 0$ for all $z \in R$,

$$|f_n(z) - f(z)| = |z/n - 0| \leq \frac{1}{n} < \varepsilon$$

for all n large enough ($n > 1/\varepsilon$), *independent of $z \in R$.*

DEFINITION   Let $\{f_n(z)\}$ ($n = 1, 2, 3, \cdots$) represent a sequence of functions defined on some region R, and suppose that for each z in R, $f_n(z) \to f(z)$ as $n \to \infty$. Then $f_n(z)$ is said to converge to $f(z)$ **uniformly on R** iff given $\varepsilon > 0$, there exists K such that if $n > K$, $|f_n(z) - f(z)| < \varepsilon$ for all z in R.

The value of this concept in one instance is that one can use it to show that if $f_n(z)$ is continuous on R for all n, and if $f_n(z)$ converges to $f(z)$ uniformly, then $f(z)$ is continuous. Here's how the proof goes:

Let $\varepsilon > 0$ and $z_0 \in R$. We know that $f_n(z) \to f(z)$ for each $z \in R$, hence by the defintion for uniform limit, there exists K such that for all $n > K$, $|f_n(z) - f(z)| < \varepsilon/3$ for all z in R. Pick any particular $n > K$. By continuity of f_n at z_0 in R, there exists $\delta > 0$ such that if $0 < |z - z_0| < \delta$, then $|f_n(z) - f_n(z_0)| < \varepsilon/3$. Now let z be any point in R such that $0 < |z - z_0| < \delta$. Thus we have

$$|f(z) - f(z_0)| = |f(z) - f_n(z) + f_n(z) - f_n(z_0) + f_n(z_0) - f(z_0)|$$

$$\leq |f(z) - f_n(z)| + |f_n(z) - f_n(z_0)| + |f_n(z_0) - f(z_0)| < \frac{\varepsilon}{3} + \frac{\varepsilon}{3} + \frac{\varepsilon}{3} = \varepsilon$$

and f is continuous at z_0, hence for all $z \in R$. $\blacksquare$

We establish two further theorems, which will eventually justify the termwise integration or differentiation of an infinite series.

THEOREM 7.1

Suppose $\{f_n(z)\}$ converges uniformly to $f(z)$ on R and that $f_n(z)$ is continuous on R for each n. Then for any curve C in R,

$$\lim_{n\to\infty} \int_C f_n(w)\,dw = \int_C f(w)\,dw = \int_C \left(\lim_{n\to\infty} f_n(w)\right) dw$$

Proof: Since $f(z)$ is continuous by the previous result, the integral of $f(z)$ exists on C. Let $\varepsilon > 0$ be given. Since $f_n(z) \to f(z)$ uniformly as $n \to \infty$, there exists K such that for $n > K$, $|f_n(z) - f(z)| < \dfrac{\varepsilon}{l(C)}$ for all z on C. Thus,

$$\left| \int_C f_n(w)\,dw - \int_C f(w)\,dw \right| = \left| \int_C [f_n(w) - f(w)]\,dw \right|$$

$$\leq \int_C |f_n(w) - f(w)|\,|dw| \leq \frac{\varepsilon}{l(C)} \cdot l(C) = \varepsilon$$

for all $n > K$. This proves that

$$\lim_{n\to\infty} \int_C f_n(w)\,dw = \int_C f(w)\,dw \quad \text{\\\\}$$

THEOREM 7.2

Suppose $\{f_n(z)\}$ converges uniformly to $f(z)$ on an open region R, and that $f_n(z)$ is differentiable for each n. Then if $\{f_n'(z)\}$ converges uniformly on R, the derivative of $f(z)$ exists, and

$$\lim_{n\to\infty} f_n'(z) = f'(z)$$

Proof: By hypothesis, assume that $f_n'(z) \to g(z)$ uniformly on R. We must prove that $f'(z)$ exists and that $g(z) = f'(z)$. Let C be any curve in R joining point c to an arbitrary point z in R. Since f_n is differentiable on an open region R, it is analytic at z. Thus f_n' is differentiable on R, hence continuous, so the integral of f_n' exists on C. By the fundamental theorem,

$$\int_C f_n'(w)\,dw = \int_c^z f_n'(w)\,dw = f_n(z) - f_n(c)$$

Also, $g(z)$ is continuous since $f_n'(z)$ is for each n. The hypothesis for Theorem **7.1** is then satisfied and we obtain

$$\int_C g(w)\,dw = \int_C \left(\lim_{n\to\infty} f_n'(w)\right) dw = \lim_{n\to\infty} \int_C f_n'(w)\,dw = \lim_{n\to\infty} [f_n(z) - f_n(c)] = f(z) - f(c)$$

This shows that the integral of $g(z)$ from c to z is independent of path in the open region R, and by Theorem **4.9** its derivative exists, with

$$\frac{d}{dz} \int_c^z g(w)\,dw = g(z)$$

Thus $f(z) - f(c) = \int_c^z g(w)\,dw$ is differentiable, and $g(z) = \dfrac{d}{dz}[f(z) - f(c)] = f'(z)$. Hence

$$\lim_{n\to\infty} f_n'(z) = g(z) = f'(z). \quad \text{\\\\}$$

The hypotheses in Theorem **7.2** are all necessary. The following examples show that you could have a uniform limit $f_n(z) \to f(z)$ as $n \to \infty$ without the existence of $f'(z)$, and even if $f'(z)$ exists, $\lim_{n\to\infty} f_n'(z)$ need not be uniform. Thus potentially simpler proofs of this theorem break down.

EXAMPLE 3

Consider the sequence defined by $f_n(z) = \sqrt{z + \dfrac{1}{n}}$, which converges to $f(z) = \sqrt{z}$. Let R be the non-negative first quadrant $x \geq 0$, $y \geq 0$ in the complex plane (Figure 7.1). Show that $f'(z)$ does not exist on R even though $f_n'(z)$ exists for each n and the limit $f_n(z) \rightarrow f(z)$ is uniform on R.

SOLUTION

To show the limit is uniform on R, observe that

$$\left| \sqrt{z + \frac{1}{n}} - \sqrt{z} \right| = \left| \frac{(\sqrt{z+1/n} - \sqrt{z})(\sqrt{z+1/n} + \sqrt{z})}{\sqrt{z+1/n} + \sqrt{z}} \right| = \left| \frac{z+1/n-z}{\sqrt{z+1/n} + \sqrt{z}} \right| = \frac{1/n}{\left| \sqrt{z+1/n} + \sqrt{z} \right|}$$

We proceed to establish $\left| \sqrt{z+c} + \sqrt{z} \right| \geq \sqrt{c}$ for real $c > 0$, where $z = x + iy$, $x \geq 0$ and $y \geq 0$. For convenience, let λ and μ be the imaginary parts of $\sqrt{z+c}$ and $\sqrt{z}$ respectively). Then by definition,

$$\left| \sqrt{z+c} + \sqrt{z} \right| = \left| \sqrt{\frac{\sqrt{(x+c)^2 + y^2} + x + c}{2}} + i\lambda + \sqrt{\frac{\sqrt{x^2 + y^2} + x}{2}} + i\mu \right|$$

$$= \left[\left(\sqrt{\frac{\sqrt{(x+c)^2 + y^2} + x + c}{2}} + \sqrt{\frac{\sqrt{x^2 + y^2} + x}{2}} \right)^2 + (\lambda + \mu)^2 \right]^{\frac{1}{2}}$$

$$\geq \sqrt{\frac{\sqrt{(x+c)^2 + y^2} + x + c}{2}} \geq \sqrt{\frac{\sqrt{c^2} + c}{2}} = \sqrt{c}$$

Thus, with $c = 1/n$,

$$\left| \sqrt{z + \frac{1}{n}} - \sqrt{z} \right| = \frac{1/n}{\left| \sqrt{z+1/n} + z \right|} \leq \frac{1/n}{\sqrt{1/n}} = \frac{1}{\sqrt{n}}$$

which shows that the zero limit does not depend on z, hence the limit is uniform on R. Also, for each n, $\dfrac{d}{dz} \sqrt{z + \dfrac{1}{n}} = 0$ if $z = 0$, or $= \frac{1}{2}(z + 1/n)^{-1/2}$ if $z \neq 0$. So, $f_n'(z)$ exists on R for all n, but $f'(z) = \dfrac{d}{dz}(\sqrt{z}) = \frac{1}{2} z^{-1/2}$ does not exist at $z = 0 \in R$. (However, R is not open as required by Theorem 7.2.)

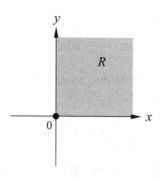

FIGURE 7.1

EXAMPLE 4

Let $f_n(z) = z^n/n$ on the open region $R : |z| < 1$. Although both limits $\lim\limits_{n \to \infty} f_n(z)$ and $\lim\limits_{n \to \infty} f_n'(z)$ exist and equal zero on R, the first converges uniformly, but the second one, $\{f_n'(z)\}$, does not.

EXAMPLE 5

Consider the sequence of functions given by $f_n(z) = \dfrac{n^2 z^2}{z^2 + n^2}$ on $|z| < 1$. Find the following limits, and verify Theorem **7.2**:

(a) $\lim\limits_{n\to\infty} f_n(z) = f(z)$

(b) $\lim\limits_{n\to\infty} \dfrac{d}{dz} f_n(z)$

SOLUTION

(a) $f_n(z) = \dfrac{n^2 z^2}{z^2 + n^2} = \dfrac{z^2}{z^2/n^2 + 1} \to \dfrac{z^2}{0+1} = z^2 = f(z)$ as $n\to\infty$,

(b) $f_n'(z) = \dfrac{2n^2 z(z^2 + n^2) - n^2 z^2(2z)}{(z^2 + n^2)^2} = \dfrac{2n^4 z}{(z^2 + n^2)^2} = \dfrac{2z}{(z^2/n^2 + 1)^2} \to 2z$

and $\lim\limits_{n\to\infty} f_n'(z_n) = 2z = \dfrac{d}{dz} z^2 = f'(z)$.

EXAMPLE 6

As in the previous example, let $f_n(z) = n^2 z^2/(z^2 + n^2)$. Verify Theorem **7.1** for the sequence of integrals for $n = 1, 2, 3, \cdots$ by evaluating and taking the limit of $\int_0^z f_n(w)dw$ as $n\to\infty$.

SOLUTION

$$\int_0^z f_n(w)\,dw = \int_0^z \frac{n^2 w^2}{w^2 + n^2}\,dw = \int_0^z \frac{n^2 w^2 + n^4 - n^4}{w^2 + n^2}\,dw = \int_0^z \left(n^2 - \frac{n^4}{w^2 + n^2}\right)dw$$

$$= \left(n^2 w - n^4 \cdot \frac{1}{n}\tan^{-1}\frac{w}{n}\right)\Bigg|_0^z = n^2 z - n^3 \tan^{-1}\frac{z}{n} - 0$$

$$= n^2 z - n^3\left(\frac{z}{n} - \frac{z^3}{3n^3} + \frac{z^5}{5n^5} - \frac{z^7}{7n^7} + \cdots\right) = \frac{z^3}{3} - \frac{z^5}{5n^2} + \frac{z^7}{7n^4} - \cdots$$

Taking the limit as $n\to\infty$, we obtain $\dfrac{z^3}{3}$. On the other hand, the limit of $f_n(z)$ was shown in Example 5 to be z^2. Thus we obtain the following verification of Theorem **7.1**

$$\lim_{n\to\infty} \int_0^z f_n(w)\,dw = \frac{z^3}{3} = \int_0^z w^2\,dw = \int_0^z \lim_{n\to\infty} f_n(w)\,dw$$

UNIFORM CONVERGENCE FOR INFINITE SERIES

A generalization of the power series introduced in Chapter 6 is provided by a series of the form

$$\sum_{n=0}^{\infty} f_n(z) = f(z)$$

where $f_n(z)$ represents a function for each n (usually assumed to be at least continuous). The set of values of z for which the series converges becomes the domain of the function $f(z)$, which must surely be continuous if each $f_n(z)$ is continuous. Do you see the necessity for a proof of this fact, if true? Do you see a connection with the results just discussed in the preceding section?

Indeed, the problem we have here is identical to that of the previous section: recall that an infinite series is by definition the *limit of a sequence*. Thus if the individual terms of a series are functions, then the concept of uniform convergence for series emerges. Strictly in terms of series, we can formulate the equivalent definition for uniform convergence.

DEFINITION   Suppose the series $\sum_{n=0}^{\infty} f_n(z)$ converges to $f(z)$ in some region R. Then the series is said to **converge uniformly** to $f(z)$ throughout R provided that for any given $\varepsilon > 0$ there exists a constant K such that for all $k > K$ the inequality $\left| \sum_{n=0}^{k} f_n(z) - f(z) \right| < \varepsilon$ holds independently of z in R.

It can be shown, for example, that each of the series

$$\sum_{n=0}^{\infty} z^n, \qquad \sum_{n=0}^{\infty} \frac{z^n}{n}, \qquad \text{and} \qquad \sum_{n=0}^{\infty} n^2 z^n,$$

converge uniformly to their limits for $|z| \leq \frac{2}{3}$. In fact, any power series converges uniformly in any closed disk concentric to and in the interior of its circle of convergence (argument provided momentarily). But a series need not converge uniformly throughout its *entire* region of convergence.

EXAMPLE 7

Show that the infinite series $\sum_{n=0}^{\infty} z^n$ is not uniformly convergent in its region of convergence.

SOLUTION

We know that this series converges to $(1 - z)^{-1}$ in the region R defined by $|z| < 1$. The partial sums of this series are, for each k and each z in R

$$s_k(z) = \sum_{n=0}^{k} f_n(z) = 1 + z + z^2 + z^3 + \cdots + z^k = \frac{1 - z^{k+1}}{1 - z}$$

Thus, letting $f_n(z) = z^n$ and $f(z) = (1 - z)^{-1}$, the requirement *if* the series were uniformly convergent would be

$$\left| \sum_{n=0}^{k} f_n(z) - f(z) \right| = \left| \sum_{n=0}^{k} f_n(z) - \frac{1}{1 - z} \right| = \left| \frac{1 - z^{k+1}}{1 - z} - \frac{1}{1 - z} \right| = \left| \frac{z^{k+1}}{1 - z} \right| < \varepsilon$$

for some $\varepsilon > 0$, for all $k >$ some K, and for all z such that $|z| < 1$. Letting k be any specific integer $> K$, this requirement is violated as $z \to 1$.

The following theorem is a result already established for sequences. Thus we conclude:

THEOREM 7.3

If a series $\sum f_n(z)$ converges uniformly to $f(z)$ on some domain R, and if each $f_n(z)$ is continuous on R, then $f(z)$ is also continuous on R.

EXAMPLE 8

The series $f(z) = z + 2z^2 + 3z^3 + 4z^4 + \cdots + nz^n + \cdots$ converges for $|z| < 1$, and is uniformly convergent for $|z| \leq r < 1$ (this fact can be established by methods to be introduced later). Each term of the series (nz^n for a fixed integer n) represents a continuous function, so by Theorem **7.3** $f(z)$ is continuous. One realizes the advantage of Theorem **7.3** when this analysis is compared to the more pedestrian approach of trying to use the definition of continuity directly; one must somehow show that for $\varepsilon \geq 0$

$$|f(z) - f(c)| = |(z + 2z^2 + 3z^3 + 4z^4 + \cdots) - (c + 2c^2 + 3c^3 + 4c^4 + \cdots)|$$
$$= |z - c + 2(z^2 - c^2) + 3(z^3 - c^3) + 4(z^4 - c^4) + \cdots| < \varepsilon$$

for z close enough to c. While each individual term in the infinite sum is "small", it is by no means evident that the entire sum of these terms is "small".

Theorems **7.1** and **7.2** have important corollaries that apply directly to infinite series, and it is not necessary to formulate separate proofs for series. Thus, the following two theorems have already been proved.

THEOREM 7.4 (Termwise Differentiation)

If a series $\sum f_n(z)$ converges uniformly to the function $f(z)$ on an open, simply-connected region R, and if the series of derivatives $f_n'(z)$ converges uniformly on R, then

(7.4)
$$f'(z) = \frac{d}{dz} \sum_{n=0}^{\infty} f_n(z) = \sum_{n=0}^{\infty} \frac{d}{dz} f_n(z)$$

THEOREM 7.5 (Termwise Integration)

If a series $\sum f_n(z)$ converges uniformly to the function $f(z)$ on a region R, and $f_n(z)$ is continuous in R for each n, then the integrals of $f_n(z)$ and $f(z)$ exist along any curve C in R, and

(7.5)
$$\int_C f(w)\, dw = \int_C \left(\sum_{n=0}^{\infty} f_n(w) \right) dw = \sum_{n=0}^{\infty} \left[\int_C f_n(w)\, dw \right]$$

These two theorems finally establish for power series the validity of either differentiating or integrating an infinite series termwise. This concept was taken for granted and used several times in Chapter 6. The hypotheses of these theorems can be checked to make sure the previous conclusions are warranted. For example Theorem **7.4** justifies a result like

$$\frac{1}{(1+z)^2} = -\frac{d}{dz}\left(\frac{1}{1+z} \right) = -\frac{d}{dz}(1 - z + z^2 - z^3 + z^4 - \cdots) = 1 - 2z + 3z^2 - 4z^3 + \cdots$$

or like

$$\ln(1+z) = \int_0^z \frac{dw}{1+w} = \int_0^z (1 - w + w^2 - w^3 + w^4 - \cdots)\, dw = z - \frac{z^2}{2} + \frac{z^3}{3} - \frac{z^4}{4} + \frac{z^5}{5} - \cdots$$

EXAMPLE 9

Use series to show that the derivative of $\cosh z$ is $\sinh z$.

SOLUTION

Using the Taylor series expansions of $\cosh z$ and $\sinh z$, we have

$$\cosh z = 1 + \frac{z^2}{2!} + \frac{z^4}{4!} + \frac{z^6}{6!} + \frac{z^8}{8!} + \cdots$$

This series converges uniformly to $\cosh z$ on the entire complex plane (see Theorem 7.7 below). Accordingly, we can differentiate the series termwise:

$$\frac{d}{dz}\cosh z = 0 + \frac{2z}{2!} + \frac{4z^3}{4!} + \frac{6z^5}{6!} + \frac{8z^7}{8!} + \cdots$$

$$= z + \frac{z^3}{3!} + \frac{z^5}{5!} + \frac{z^7}{7!} + \cdots = \sinh z$$

We now move on to a very important theorem on uniform convergence known as the *Weierstrass M-test*. First, an observation will make the definition of uniform convergence for series easier to work with. Instead of looking at the entire partial sum s_n in the definition of uniform convergence, suppose we can show that, given $\varepsilon > 0$, there exists K such that for all $k > K$ and $p > 0$,

$$\left| \sum_{n=k}^{k+p} f_n(z) \right| < \varepsilon$$

for all $z \in R$. Notice that by manipulating with partial sums and using the triangle inequality:

$$\left| \sum_{n=0}^{k-1} f_n(z) - f(z) \right| = \left| -\sum_{n=k}^{k+p} f_n(z) + \sum_{n=0}^{k+p} f_n(z) - f(z) \right|$$

$$\leq \left| \sum_{n=k}^{k+p} f_n(z) \right| + \left| \sum_{n=0}^{k+p} f_n(z) - f(z) \right|$$

$$< \varepsilon + \left| \sum_{n=0}^{k+p} f_n(z) - f(z) \right|$$

Now let $p \to \infty$; the term in absolute values converges to zero. Hence

$$\left| \sum_{n=0}^{k-1} f_n(z) - f(z) \right| \leq \varepsilon$$

and it follows that the series converges to $f(z)$ uniformly on R. This establishes the following.

THEOREM 7.6

The series $\sum f_n(z)$ converges uniformly on some domain R iff given $\varepsilon > 0$ there exists K such that if $k > K$ and $p > 0$,

$$\left| \sum_{n=k}^{k+p} f_n(z) \right| < \varepsilon$$

for all z in R.

(The proof of the converse will be left to the reader; start with the tail of the series and essentially use a reversal of the above steps and prove that the tail is $\leq \varepsilon$.)

Another major theorem of this section uses the fact that if a series of non-negative *real numbers* converges, then the partial sums converge, leaving the difference (the sequence of the tails of the series) to converge to zero. That is, given $\varepsilon > 0$, there exists K such that for all $k > K$, the sum from the kth term on is less than ε. Conversely, if all the tails of a series beyond a certain point are arbitrarily small, then the series converges (this was explicitly shown in Problems 48–49, Chapter 6).

THEOREM 7.7 (The Weierstrass M Test)

Suppose that real numbers $M_1, M_2, \cdots, M_n, \cdots$ can be found such that the series $\sum M_n$ converges and such that $|f_n(z)| \leq M_n$ for each n and for all z in R. Then the series $\sum f_n(z)$ converges uniformly on R.

Proof: By the remark preceding the theorem, given $\varepsilon > 0$ there exists K such that if $k > K$ and $p > 0$,

$$\sum_{n=k}^{k+p} M_n < \varepsilon$$

But, by hypothesis

$$\left| \sum_{n=k}^{k+p} f_n(z) \right| \leq \sum_{n=k}^{k+p} |f_n(z)| \leq \sum_{n=k}^{k+p} M_n < \varepsilon$$

for all z in R. By Theorem 7.6, the series $\sum f_n(z)$ converges uniformly on R. ⧹

EXAMPLE 10

Show that the series $\sum 3^n z^n$ converges uniformly for $|z| \leq \frac{1}{4}$.

SOLUTION

Since $|3^n z^n| \leq 3^n (\frac{1}{4})^n = (\frac{3}{4})^n$, let $M_n = (\frac{3}{4})^n$. Consider the series $\sum (\frac{3}{4})^n$, which converges as a geometric serie. The Weierstrass M test applies, showing uniform convergence for $|z| \leq \frac{1}{4}$. (The series can be seen to converge to the continuous function $f(z) = (1 - 3z)^{-1}$ for $|z| < \frac{1}{3}$, but the series is not uniformly convergent on $|z| < \frac{1}{3}$, only the part $|z| \leq \frac{1}{4}$ (or, as can be shown, for $|z| \leq s$ for any positive $s < \frac{1}{3}$). The next theorem provides a general result concerning the conclusion we obtained here.

THEOREM 7.8

If the radius of convergence of the power series $\sum a_n z^n$ is $r > 0$, then it is uniformly convergent for all z within any closed disk $|z| \leq s$ strictly inside the open disk $|z| < r$ ($s < r$).

Proof: We first show that the series $\sum |a_n| s^n$ converges. Let z_1 be any nonzero point inside the circle $|z| < r$ and outside the circle $|z| < s$, as shown in Figure 7.2. Hence for all large enough n. Since $\sum a_n z_1^n$ converges), $|a_n z_1^n| < 1$ or $|a_n| < 1/|z_1^n|$ for all sufficiently large n. Therefore, by multiplying by s^n,

$$|a_n| s^n < \frac{s^n}{|z_1|^n} = t^n$$

where $t = s/|z_1| < 1$. Thus, $\sum t^n$ converges. By the comparison test (for real series), $\sum |a_n| s^n$

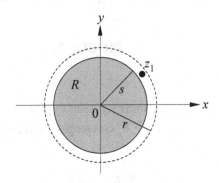

FIGURE 7.2

converges. Now set $M_n = |a_n|s^n$ for each n. If $z \in R$ (hence $|z| \le s$), then $|a_n z^n| \le |a_n|s^n = M_n$, and by the Weierstrass M test, the given series is uniformly convergent on R. $\backslash\!\backslash$

Now we come to the question posed by the following conjecture:

$$\lim_{z \to c} \sum_{n=0}^{\infty} f_n(z) = \sum_{n=0}^{\infty} f_n(c)$$

This cannot be true in general since we know, for example, that as $z \to 1^-$ the convergent series $\sum_{n=0}^{\infty} \dfrac{z^n}{n}$ $(|z| < 1)$ approaches the harmonic series and does not converge. Suppose, then, that we require the series *evaluated at c* to converge; can anything be said? Indeed, a positive result can be obtained (*for power series only*). It is due to another prominent pioneer in complex analysis, N. H. Abel (1802–1829). The interesting thing about this theorem is that no form of uniform convergence is required for the given series.

THEOREM 7.9 (Boundary Theorem of Abel)

Suppose that the infinite series $\sum a_n z^n$ converges for all z in the open region R given by $|z| < r$, and that for some z_0 on the boundary of R, $\sum a_n z_0^n$ converges (Figure 7.3). If S is the geometric radius joining 0 and z_0, then as $z \to z_0$ on S,

(7.9)
$$\lim_{z \to z_0} \sum a_n z^n = \sum a_n z_0^n$$

Proof: Segment S can be parameterized as $z = t z_0$ $(0 \le t \le 1)$. Hence we must establish the limit

$$\lim_{t \to 1^-} \sum_{n=0}^{\infty} a_n (t z_0)^n = \sum_{n=0}^{\infty} a_n z_0^{\,n}$$

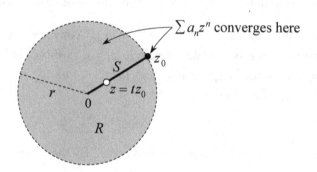

FIGURE 7.3

To simplify notation, let $a_n z_0^n = b_n$; the limit we want to prove becomes simply $\lim \sum b_n t^n = \sum b_n$ as $t \to 1$. By hypothesis, the series $\sum b_n$ converges, so for each $\varepsilon > 0$ there exists K such that

$$\left| \sum_{n=k}^{N} b_n \right| < \varepsilon/2$$

for all $k > K$ and any $N > k$. Note that for each m such that $k \leq m \leq N$, we also have

$$\left| \sum_{n=m}^{N} b_n \right| < \varepsilon/2$$

For convenience, define the partial sums

$$s_k = \sum_{n=k}^{N} b_n, \qquad s_{k+1} = \sum_{n=k+1}^{N} b_n, \qquad s_{k+2} = \sum_{n=k+2}^{N} b_n, \qquad \cdots, \qquad s_N = b_N$$

By the above, it follows that $|s_m| < \varepsilon/2$ for $m = k, k+1, k+2, \cdots, N$. Also, note that

$$b_k = s_k - s_{k+1}, \quad b_{k+1} = s_{k+1} - s_{k+2}, \quad b_{k+2} = s_{k+2} - s_{k+3}, \quad \cdots, \quad b_{N-1} = s_{N-1} - s_N, \quad b_N = s_N$$

Now with t any particular real number on $(0, 1]$,

$$\left| \sum_{n=k}^{N} b_n t^n \right| = \left| b_k t^k + b_{k+1} t^{k+1} + b_{k+2} t^{k+2} + \cdots + b_N t^N \right|$$

$$= \left| (s_k - s_{k+1}) t^k + (s_{k+1} - s_{k+2}) t^{k+1} + (s_{k+2} - s_{k+3}) t^{k+2} + \cdots + (s_{N-1} - s_N) t^{N-1} + s_N t^N \right|$$

$$= \left| s_k t^k + s_{k+1}(t^{k+1} - t^k) + s_{k+2}(t^{k+2} - t^{k+1}) + \cdots + s_N(t^N - t^{N-1}) \right|$$

$$\leq |s_k| \, |t^k| + |s_{k+1}| \, |t^{k+1} - t^k| + |s_{k+2}| \, |t^{k+2} - t^{k+1}| + \cdots + |s_N| \, |t^N - t^{N-1}|$$

$$< (\varepsilon/2) \, [t^k + (t^k - t^{k+1}) + (t^{k+1} - t^{k+2}) + \cdots + (t^{N-1} - t^N)]$$

$$< (\varepsilon/2) \cdot 2 t^k \leq \varepsilon$$

Hence, by Theorem 7.6, $\sum b_n t^n$ is uniformly convergent on the closed segment S, and by Theorem **7.3** it defines a continuous function $g(t)$ on S, where $g(1) = \sum b_n$. Thus, as $z \to z_0$ on S, then $t \to 1$ and we reach the desired conclusion

$$\lim_{z \to z_0} \sum a_n z^n = \lim_{t \to 1^-} \sum b_n t^n = \lim_{t \to 1^-} g(t) = g(1) = \sum b_n = \sum a_n z_0^n . \quad \diagdown\!\!\!\diagdown$$

NOTE: It might appear that Abel's theorem is much ado about nothing, particularly since it seems that one should be able to easily conclude that $\lim_{t \to 1} \sum b_n t^n = \sum b_n$ without using the partial sums s_m. However, this would involve a double limit which is characteristically problematic in mathematics. Furthermore, a unique example due to G.H. Hardy and J.C. Littlewood mentioned in (Titchmarsh, 1975, page 231), shows that if the segment S is replaced by a curve joining 0 and z_0 that is *tangent to the circle of convergence*, the above conclusion *fails!* $\diagdown\!\!\!\diagdown$

COROLLARY

Suppose that the series $\sum a_n z^n$ converges to a function $f(z)$ on the open disk $|z| < r$, that $\sum a_n z_0^n$ converges for $|z_0| = r$, and that $f(z)$ is continuous at $z = z_0$. Then

(7.10)
$$\sum a_n z_0^n = f(z_0).$$

Proof: The question is whether the value of the series at $z = z_0$ actually equals $f(z_0)$. Restricting the values of z to those on the radius S (defined above), Abel's theorem showed that $\sum a_n z_0^n$ equals the limit of $\sum a_n z^n$ as $z \to z_0$ on S. But since $f(z)$ is continuous on S, we obtain

$$\sum a_n z_0^n = \lim_{z \to z_0} \sum a_n z^n = \lim_{z \to z_0} f(z) = f(z_0) \quad \diagdown\!\!\!\diagdown$$

Abel's theorem has immediate applications, not only for complex variables, but for problems often encountered in calculus. For example, recall the calculus result

$$1 - \frac{1}{2} + \frac{1}{3} - \frac{1}{4} + \frac{1}{5} - \frac{1}{6} + - \cdots = \ln 2$$

which, if actually proved, requires the use of the remainder term in the Taylor series expansion of the function $\ln(1 + x)$ for $|x| < 1$. One must prove that at $x = 1$, this remainder term has limit zero as $n \to \infty$ (a detail that is often omitted). But Abel's theorem (and the corollary) provides the appropriate justification: since $\ln(1 + x)$ is continuous at $x = 1$ and the alternating harmonic series is convergent there, then the substitution of 1 for x in the series for $\ln(1 + x)$ is permitted and you get the above result. This would also justify going from the Taylor series expansion

$$\tan^{-1} x = x - \frac{x^3}{3} + \frac{x^5}{5} - \frac{x^7}{7} + - \cdots + \frac{(-x)^{2n+1}}{2n+1} + \cdots$$

(valid for $|x| < 1$) to the result

$$\tan^{-1} 1 = \frac{\pi}{4} = 1 - \frac{1}{3} + \frac{1}{5} - \frac{1}{7} + - \cdots + \frac{(-1)^{2n+1}}{2n+1} + \cdots$$

This formula probably appeared in your calculus text, without adequate proof. Abel's theorem also justifies the method we used in Examples 9–11 in Chapter 6.

Here is a more substantial application of Abel's theorem and (7.10) to obtain the exact value of a series of real numbers that, at face value, appears to be quite difficult.

EXAMPLE 11

Find the value of the series

$$1 - \left(\frac{1}{2} \cdot \frac{1}{3} \right) + \left(\frac{1}{2} \cdot \frac{3}{4} \cdot \frac{1}{5} \right) - \left(\frac{1}{2} \cdot \frac{3}{4} \cdot \frac{5}{6} \cdot \frac{1}{7} \right) + \left(\frac{1}{2} \cdot \frac{3}{4} \cdot \frac{5}{6} \cdot \frac{7}{8} \cdot \frac{1}{9} \right) + - \cdots$$

SOLUTION

Since this series resembles the known real series for $\sin^{-1} x$ at $x = -1$, we are led to look at the complex series for $\sin^{-1} z$ for $|z| < 1$. As developed in Problem 39, Chapter 6, the result is, for $|z| < 1$,

$$\sin^{-1} z = z + \sum_{n=1}^{\infty} \left(\frac{1 \cdot 3 \cdot 5 \cdots (2n-1)}{2 \cdot 4 \cdot 6 \, \cdots \, (2n)} \right) \frac{z^{2n+1}}{2n+1}$$

$$= z + \frac{1}{2} \cdot \frac{z^3}{3} + \frac{1 \cdot 3}{2 \cdot 4} \cdot \frac{z^5}{5} + \frac{1 \cdot 3 \cdot 5}{2 \cdot 4 \cdot 6} \cdot \frac{z^7}{7} + \cdots + \left(\frac{1 \cdot 3 \cdot 5 \cdots (2n-1)}{2 \cdot 4 \cdot 6 \, \cdots \, (2n)} \right) \frac{z^{2n+1}}{2n+1} + \cdots$$

We are going to substitute $z = i$ on both sides, but in order to make this legitimate, we must use Abel's theorem, and this requires the series to converge for $z = i$. Thus first it must be shown that

$$= i + \frac{1}{2} \cdot \frac{i^3}{3} + \frac{1 \cdot 3}{2 \cdot 4} \cdot \frac{i^5}{5} + \frac{1 \cdot 3 \cdot 5}{2 \cdot 4 \cdot 6} \cdot \frac{i^7}{7} + \cdots + \left(\frac{1 \cdot 3 \cdot 5 \cdots (2n-1)}{2 \cdot 4 \cdot 6 \, \cdots \, (2n)} \right) \frac{i^{2n+1}}{2n+1}$$

$$= i - \frac{1}{2} \cdot \frac{i}{3} + \frac{1 \cdot 3}{2 \cdot 4} \cdot \frac{i}{5} - \frac{1 \cdot 3 \cdot 5}{2 \cdot 4 \cdot 6} \cdot \frac{i}{7} + \cdots + \left(\frac{1 \cdot 3 \cdot 5 \cdots (2n-1)}{2 \cdot 4 \cdot 6 \, \cdots \, (2n)} \right) \frac{(-1)^n i}{2n+1} + \cdots$$

converges. This is i times the given alternating series, which, as a real series, converges if the nth term converges to zero and its terms in absolute value are non-increasing. Observe that

$$\frac{1}{2} \cdot \frac{3}{4} \cdot \frac{5}{6} \cdots \frac{2n-1}{2n} \cdot \frac{1}{2n+1} < \frac{1}{2} \cdot \frac{3}{4} \cdot \frac{5}{6} \cdots \frac{2n-3}{2n-2} \cdot \frac{1}{2n-1} < \frac{1}{2n-1}$$

So by the alternating series theorem in calculus, this series converges. Multiplying by i does not affect convergence, hence the series representing $\sin^{-1}z$ converges at i. By (7.10) this series converges to $\sin^{-1}i$, and the given (real) series converges to the value $-i\sin^{-1}i$. Thus we seek the value of $\sin^{-1}i$, that is, we need a complex number c such that $\sin c = i$. With $c = a + bi$, by definition

$$\sin(a+bi) = \sin a \cosh b + i\cos a \sinh b = i \quad \Rightarrow \quad \sin a \cosh b = 0 \text{ and } \cos a \sinh b = 1$$

Since $\cosh b \neq 0$, $\sin a = 0$ and $a = k\pi$ for k an integer. The other equation then becomes, by substitution, $\cos k\pi \sinh b = 1$, which requires k to be an even integer and $\sinh b = 1$. That is, for some integer k,

$$\sin^{-1}i = c = 2k\pi + i\sinh^{-1}1$$

and the value of the given series must be $-i(2k\pi + i\sinh^{-1}1) = -2k\pi i + \sinh^{-1}1$. Since the series is real, $k = 0$. We have then found that the given series has the value

$$\sinh^{-1}1 \equiv \ln(1 + \sqrt{2}).$$

PROBLEMS

1. Show that $\lim\limits_{n\to\infty} \dfrac{nz}{n+1} = z$ for all fixed z, and determine for what region the limit is uniform.

2. For all z in the entire complex plane R, the limit of e^{-nz} equals zero as $n\to\infty$.
   (a) Is this limit uniform in R?
   (b) Find a region for which the limit is uniform.

3. Evaluate the following limits if they exist:
   (a) $\lim\limits_{n\to\infty} \dfrac{1}{n}\sqrt{1+n^2z}$
   (b) $\lim\limits_{n\to\infty} \dfrac{d}{dz}\left(\dfrac{1}{n}\sqrt{1+n^2z}\right)$
   (c) $\lim\limits_{n\to\infty} \int_0^z \dfrac{1}{n}\sqrt{1+n^2w}\,dw$

4. Show that $\lim\limits_{n\to\infty} \dfrac{n+1}{nz} = \dfrac{1}{z}$ uniformly on the region R consisting of the open annular ring $\frac{1}{10} < |z| < 100$. Discuss Theorems 7.1 and 7.2 with regard to
   (a) $\lim\limits_{n\to\infty} \dfrac{d}{dz}\left(\dfrac{n+1}{nz}\right)$
   (b) $\lim\limits_{n\to\infty} \int_1^z \dfrac{n+1}{nw}\,dw$

5. Evaluate the following limits, and discuss Theorems 7.1 and 7.2 with regard to (b) and (c):
   (a) $\lim\limits_{n\to\infty} nz\sin\dfrac{z^2}{n}$ [**Hint:** $nz\sin\dfrac{z^2}{n} = \dfrac{\sin\frac{z^2}{n}}{\frac{z^2}{n}}\cdot z^3 = \dfrac{\sin\theta}{\theta}z^3.$]
   (b) $\lim\limits_{n\to\infty} \dfrac{d}{dz}\left(nz\sin\dfrac{z^2}{n}\right)$
   (c) $\lim\limits_{n\to\infty} \int_0^z nw\sin\dfrac{w^2}{n}\,dw$

6. Differentiate the series for $\sin z$ to derive the series for $\cos z$.

7. Differentiate the series for $\ln(1+z)$ to derive the geometric series for $(1+z)^{-1}$.

8. Derive the Taylor series for $\tan^{-1}z$ by using the geometric series $1 - z^2 + z^4 - z^6 + \cdots$ and integrating term-wise. Justify the procedure and determine for what region the series is valid. (Recall Theorem 6.11, Chapter 6, which shows that the series obtained is the Taylor series.)

9. Using the result of Problem 8 and the corollary to Abel's theorem 7.10 derive the following expression for π, and justify your answer:

$$4 - \frac{4}{3} + \frac{4}{5} - \frac{4}{7} + \frac{4}{9} - \cdots$$

Using the first twenty terms, what estimate does this series produce for π? (Recall the alternating series theorem from calculus that predicts an error between the series and the nth partial sum to be at most the value of the $(n+1)$st term of the series; compare this with the result you obtained for π.)

10. Use termwise integration to establish the following series expansion

$$\frac{z^2}{1 \cdot 2} - \frac{z^3}{2 \cdot 3} + \frac{z^4}{3 \cdot 4} - \frac{z^5}{4 \cdot 5} + \cdots + \frac{(-z)^{n+1}}{n(n+1)} + \cdots = (z+1)\left[\ln(z+1) - 1\right]$$

and justify each step. Find the radius of convergence.

11. Using the series of Problem 10, justify the following numerical result:

$$\frac{1}{6} - \frac{1}{20} + \frac{1}{42} - \frac{1}{72} + \cdots + \frac{(-1)^{n+1}}{2n(2n+1)} + \cdots = \ln\sqrt{2} + \frac{\pi}{4} - 1$$

12. To what algebraic function does the following series converge?

$$z + 2z^2 + 3z^3 + 4z^4 + \cdots + nz^n + \cdots$$

13. Provide the details needed in order to establish the result of Problem 39(c), Chapter 6.

14. Provide the details needed in order to establish the result of Problems 41 and 42, Chapter 6.

15. **Uniform Continuity.** The concept of uniform continuity for a function on a region R is an important concept for analysis. The definition is:

> A function $f(z)$ is said to be **uniformly continuous on R** iff for each $\varepsilon > 0$ there exists $\delta > 0$ such that if $|z - w| < \delta$ for any two points z and w in R, then $|f(z) - f(w)| < \varepsilon$.

First extend the Heine-Borel theorem to any closed and bounded region. Then prove that a continuous function on a closed and bounded region R is uniformly continuous on R. [**Hint:** For each $\varepsilon > 0$ and $z_0 \in R$ there exists a neighborhood $N(z_0, \delta)$ such that for all $z \in N(z_0, \delta)$, $|f(z) - f(z_0)| < \varepsilon/2$. The open neighborhoods $N(z_0, \delta/2)$ cover R. By the Heine-Borel theorem, a finite collection $N(z_1, \delta_1/2)$, $N(z_2, \delta_2/2)$, $N(z_3, \delta_3/2)$, $\cdots$, $N(z_n, \delta_n/2)$ covers R; take $\delta = \frac{1}{2}\min\{\delta_1, \delta_2, \delta_3, \cdots, \delta_n\}$.]

16. **The Lipschitz Condition.** Another important concept in analysis is the *Lipschitz condition* for a function: There exists a real constant k such that for all z and w on R, $|f(z) - f(w)| \le k|z - w|$. This condition is stronger than uniform continuity on R (it implies uniform continuity) but weaker than differentiability on R (it does not imply differentiability). [For example, $|\bar{z} - \bar{w}| = |z - w|$ but $\bar{z}$ is not differentiable.] Show that:

(a) If $f(z)$ satisfies the Lipschitz condition on R, then $f(z)$ is uniformly continuous on R.

(b) If $f(z) \equiv f(x + iy) = 2x + 3iy$ then $|f(z) - f(w)| \le 3|z - w|$ but $f(z)$ is not differentiable.

(c) If $f(z)$ is differentiable on a closed and bounded region R, then $f(z)$ satisfies the Lipschitz condition on R. [**Hint:** Use the approximation theorem and the fact that $f'(z)$ is bounded on R.]

LAURENT SERIES AND

RESIDUE THEORY

A simple algebraic maneuver will convert **(6.12)** to an identity that will lead to a more general type of power series expansion $\sum a_n z^n$, where instead of the exponent n being a positive integer and ranging from 0 to ∞, it can be a negative integer as well. It is not difficult to observe valid power series that have negative exponents. To see this, we begin by considering a few examples before tackling the general case.

POWER SERIES HAVING NEGATIVE EXPONENTS

Suppose one multiplies a power series in z^n by a simple factor, such as z^{-p}. For example, if we multiply the geometric series $\sum z^n$ by z^{-3} we obtain

$$z^{-3} + z^{-2} + z^{-1} + 1 + z + z^2 + z^3 + z^4 + z^5 + \cdots \qquad (0 < |z| < 1)$$

which has both positive and negative exponents. Presumably, this particular series converges to the function

$$\frac{1}{z^3} \cdot \frac{1}{1-z} = \frac{1}{z^3 - z^4}$$

having singular points at $z = 0, 1$. The region of convergence is the deleted disk defined by

$$0 < |z| < 1$$

(illustrated in Figure 8.1). A power series in z having both positive and negative exponents is called a **Laurent series**.

The next example involves a series that converges for the region $1 < |z| < 2$, illustrated in Figure 8.2).

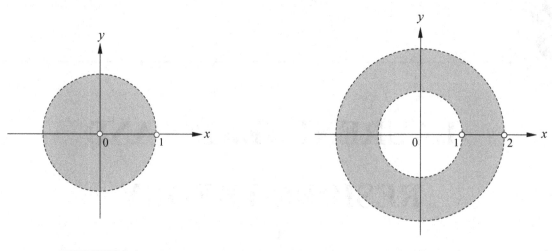

FIGURE 8.1   **FIGURE 8.2**

EXAMPLE 1

Use algebra and the formula for geometric series to find a power series representation for the function

$$f(z) = \frac{-z}{z^2 - 3z + 2}$$

that contains both positive and negative exponents, and find the region of convergence.

SOLUTION

Recall the method of partial fractions from calculus. The given function must equal

$$\frac{-z}{(z-1)(z-2)} = \frac{A}{z-1} + \frac{B}{z-2}$$

for certain constants A and B. Thus, by multiplying both sides by $(z-1)(z-2)$, we obtain the identity

$$-z = A(z-2) + B(z-1)$$

If we let $z = 1, 2$, we obtain

$$-1 = A(-1) + B \cdot 0 \qquad \Rightarrow \qquad A = 1$$
$$-2 = A \cdot 0 + B(1) \qquad \Rightarrow \qquad B = -2$$

Therefore,

$$f(z) = \frac{1}{z-1} + \frac{-2}{z-2} = \frac{1}{z} \cdot \frac{1}{1-(1/z)} + \frac{1}{1-(z/2)}$$

Using the geometric series, one with z replaced by $1/z$, the other with z replaced by $z/2$, we obtain:

$$f(z) = \cdots + \frac{1}{z^n} + \cdots + \frac{1}{z^4} + \frac{1}{z^3} + \frac{1}{z^2} + \frac{1}{z} + 1 + \frac{z}{2} + \frac{z^2}{4} + \frac{z^3}{8} + \frac{z^4}{16} + \cdots + \frac{z^n}{2^n} + \cdots$$

In order for the first half of the series (with negative exponents) to converge, we must have $|z| > 1$), and for the series with positive exponents to converge, we must have $|z/2| < 1$ (or $|z| < 2$). Altogether, the combined series converges provided

$$1 < |z| < 2.$$

This type of region in the complex plane is an (open) **annular ring** having a boundary consisting of concentric circles $|z| = 1$ (inner boundary) and $|z| = 2$ (outer boundary). See Figure 8.2.

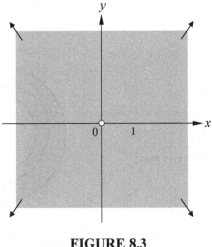

FIGURE 8.3

EXAMPLE 2

To what function in closed form does the series $\displaystyle\sum_{n=-\infty}^{\infty} \frac{z^n}{|n|!}$ converge, and what is the region of convergence?

SOLUTION

The summation symbol having limits from $-\infty$ to $+\infty$ is taken to mean the *sum* of two separate series, one with all negative exponents, the other with all positive or zero exponents. That is,

$$\sum_{n=-1}^{-\infty} \frac{z^n}{|n|!} + \sum_{n=0}^{\infty} \frac{z^n}{|n|!} = \sum_{n=1}^{\infty} \frac{1}{n!z^n} + \sum_{n=0}^{\infty} \frac{z^n}{n!}$$

Observing the known series expansion $e^w = \sum w^n/n!$) we obtain for the first series the function $e^w - 1$ (where $w = 1/z$) and e^z itself for the second series. Thus,

$$\sum_{n=1}^{\infty} \frac{1}{n!}\left(\frac{1}{z}\right)^n = e^{1/z} - 1 \qquad \text{and} \qquad \sum_{n=0}^{\infty} \frac{z^n}{n!} = e^z$$

Hence we conclude that

$$\sum_{n=-\infty}^{\infty} \frac{z^n}{|n|!} = e^{1/z} + e^z - 1$$

The region of convergence is that for the combined series, which converges for all z, *except $z = 0$*. Hence, the region of convergence is the deleted plane, $0 < |z| < \infty$, as illustrated in Figure 8.3.

Note that the regions in Figures 8.1 and 8.3 can be regarded as a type of annular ring, qualified as *virtual annular rings*. In Figures 8.1 and 8.3 the inner circle has zero radius, and in Figure 8.3 the outer circle has infinite radius.

LAURENT SERIES FOR FUNCTIONS HAVING SINGULAR POINTS

Consider an open annular ring centered at $z = c$ (Figure 8.4), defined as the open set R of all z such that $r_2 < |z - c| < r_1$ where the boundary of R consists of two concentric circles $C_1 : |z - c| = r_1$ and $C_2 : |z - c| = r_2$ with center c. Suppose $f(z)$ is any continuous function on this annular ring and its boundary. Then as w varies on either C_1 or C_2, the integrals $f(w)$ over C_1 and C_2 exist. As in Taylor's theorem in Chapter 6, let

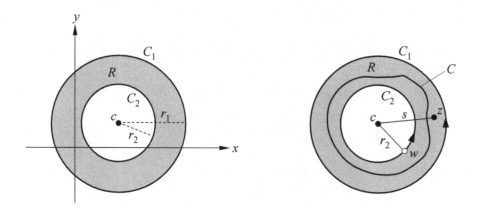

FIGURE 8.4

z be any particular (fixed) point in R (thus, if $|z - c| = s$ as in Figure 8.4, $r_2 < s < r_1$). Then for each fixed z, the integrals of $f(w)/(w - z)$ over C_1 and C_2 also exist, as do the integrals of $f(w)(w - c)^n$ where n is, more generally, any integer (positive or negative). This set-up is identical to that used for Taylor's theorem, except for the additional inner boundary C_2. The integral over C_1 was already established in **(6.13)**. We concentrate here on the integral over C_2. (The direction of integration is assumed to be counterclockwise on both circles.)

The starting point is the relation **(6.12)**, which we repeat here for convenience, including its derivation as a brief review:

$$\frac{1}{w - z} = \frac{1}{w - c}(1 + W + W^2 + W^3 + \cdots + W^n + \cdots$$

where $W = \dfrac{z - c}{w - c}$, and the geometric series $\sum W^n$ converges. Thus,

[6.12] $\quad \dfrac{1}{w - z} = \dfrac{1}{w - c} + \dfrac{z - c}{(w - c)^2} + \dfrac{(z - c)^2}{(w - c)^3} + \dfrac{(z - c)^3}{(w - c)^4} + \cdots + \dfrac{(z - c)^n}{(w - c)^{n+1}} + \cdots \qquad (W = \dfrac{z - c}{w - c})$

Suppose we simply switch w and z. The result is the infinite series of inverted terms

(8.1) $\quad -\dfrac{1}{w - z} = \dfrac{1}{z - c} + \dfrac{w - c}{(z - c)^2} + \dfrac{(w - c)^2}{(z - c)^3} + \dfrac{(w - c)^3}{(z - c)^4} + \cdots + \dfrac{(w - c)^n}{(z - c)^{n+1}} + \cdots \qquad (W = \dfrac{w - c}{z - c})$

This series converges since it is the constant $(z - c)^{-1}$ times the geometric series $\sum W^n$ where $|W| = \left|\dfrac{w - c}{z - c}\right| = \dfrac{r_2}{s} < 1$. (Thus it converges *uniformly* on any closed region $|W| \le p < r_2/s$.) Now multiply both sides by $f(w)/2\pi i$ and integrate termwise over C_2, in the positive direction (Figure 8.4). The result is

$$-\frac{1}{2\pi i}\int_{C_2} \frac{f(w)}{w - z}\,dw = \frac{1}{2\pi i}\int_{C_2} \frac{f(w)}{z - c}\,dw + \frac{1}{2\pi i}\int_{C_2} \frac{f(w)(w - c)}{(z - c)^2}\,dw$$

$$+ \frac{1}{2\pi i}\int_{C_2} \frac{f(w)(w - c)^2}{(z - c)^3} + \cdots + \frac{1}{2\pi i}\int_{C_2} \frac{f(w)(w - c)^n}{(z - c)^{n+1}} + \cdots$$

Since $z - c$ is constant, the above equation may be written more conveniently as

$$(8.2) \qquad -\frac{1}{2\pi i}\int_{C_2}\frac{f(w)}{w-z}dw = \frac{a_{-1}}{z-c}+\frac{a_{-2}}{(z-c)^2}+\frac{a_{-3}}{(z-c)^3}+\cdots = \sum_{n=1}^{\infty}\frac{a_{-n}}{(z-c)^n}$$

$$\left[\text{where } a_{-n}=\frac{1}{2\pi i}\int_{C_2}f(w)(w-c)^{n-1}dw\right]$$

The integral of $f(z)/(w-z)$ over C_1 is given by (6.13):

$$[6.13] \qquad \frac{1}{2\pi i}\int_{C_1}\frac{f(w)}{w-z}dw = \sum_{n=0}^{\infty}a_n(z-c)^n \quad \text{where} \quad a_n=\frac{1}{2\pi i}\int_{C_1}\frac{f(w)}{(w-c)^{n+1}}dw$$

Finally, suppose that $f(z)$ is analytic on $R\cup C$, where $C=C_1+(-C_2)$ is the boundary of the multiply-connected region R. This enables us to use Cauchy's integral formula as applied to $R\cup C$. Hence

$$2\pi if(z) = \int_C\frac{f(w)}{w-z}dw = \int_{C_1}\frac{f(w)}{w-z}dw+\int_{-C_2}\frac{f(w)}{w-z}dw = \int_{C_1}\frac{f(w)}{w-z}dw-\int_{C_2}\frac{f(w)}{w-z}dw$$

or, dividing by the constant $2\pi i$, and using (8.2)

$$(8.3) \qquad f(z) = \frac{1}{2\pi i}\int_{C_1}\frac{f(w)}{w-z}dw-\frac{1}{2\pi i}\int_{C_2}\frac{f(w)}{w-z}dw = \sum_{n=0}^{\infty}a_n(z-c)^n + \sum_{n=1}^{\infty}a_{-n}(z-c)^{-n}$$

where a_n and a_{-n} are the integrals given in (6.13) and (8.2). This analysis proves:

THEOREM 8.4 (Laurent's Theorem)
Let $f(z)$ be analytic on the closed annular ring R bounded by the outer circle C_1 and inner circle C_2. Then for any z in the interior of R,

$$(8.4) \qquad f(z) = \sum_{n=-\infty}^{\infty}a_n(z-c)^n \quad \text{where} \quad a_n=\frac{1}{2\pi i}\int_{C_k}\frac{f(w)}{(w-c)^{n+1}}dw$$

with $k=1$ when $n\geq 0$ and $k=2$ when $n<0$.

The two integrals for $k=1, 2$ in the theorem can be replaced by a single integral over any curve C that lies within the annular ring (see Figure 8.4). This makes it more convenient to apply Laurent's theorem. For w in R, $w-c\neq 0$, hence $f(w)(w-c)^m$ is analytic for any integer m. By Cauchy's theorem, the integrals given for a_m in (8.4) are independent of path. So the curve C can replace both C_1 and C_2 in (8.4). This proves:

COROLLARY
With the same hypothesis as the theorem, let C be any simple closed curve in R oriented counterclockwise, whose exterior contains C_1 and whose interior contains C_2. Then $f(z)$ has a Laurent series representation valid for z in R of the form

$$(8.5) \qquad f(z) = \sum_{n=-\infty}^{\infty}a_n(z-c)^n \quad \text{where} \quad a_n=\frac{1}{2\pi i}\int_C\frac{f(w)}{(w-c)^{n+1}}dw$$

It is important to recognize (8.5) as the series

$$\cdots+\frac{a_{-n}}{(z-c)^n}+\cdots+\frac{a_{-2}}{(z-c)^2}+\frac{a_{-1}}{z-c}+a_0+a_1(z-c)+a_2(z-c)^2+\cdots+a_n(z-c)^n+\cdots$$

An important concept involves singularities at $z=c$ in the series (8.5). There are two types of singularities in Laurent series: (1) When the series has an infinite number of terms with negative powers, and (2) when only a finite number of such terms are present. In the first case, the singularity is called an **essential singularity**, and in the second, a **pole**. It is said to be a **pole of order** k if k is the first index for which a negative power occurs. This is when (8.5) takes on the form

$$f(z) = \frac{a_{-k}}{(z-c)^k} + \frac{a_{-k+1}}{(z-c)^{k-1}} + \cdots + \frac{a_{-2}}{(z-c)^2} + \frac{a_{-1}}{z-c} + a_0 + a_1(z-c) + a_2(z-c)^2 + \cdots + a_n(z-c)^n + \cdots$$

If $k = 1$, the singularity c is called a **simple pole**. This case occurs quite often, and leads to a simple procedure for finding a_{-1}.

As we will see in the next few examples, one almost never uses the integral formulas in **(8.5)** to compute Laurent series. By and large, one uses simple algebra and the formula for geometric series to find the coefficients a_n. This makes use of a theorem, analogous to the uniqueness theorem for Taylor series (Theorem 6.16), that if a series converges to a function $f(z)$ in some region R, then it must be the Laurent series for $f(z)$ over R. (The proof of this theorem is substantially the same as that for Theorem 6.16.)

EXAMPLE 3

Expand the function $f(z) = \dfrac{1}{z(z-3)}$ about its singular points and state the regions of validity in each case: **(a)** about $z = 0$ and **(b)** about $z = 3$

SOLUTION

(a) First, using the method of partial fractions,

$$\frac{1}{z(z-3)} = \frac{A}{z} + \frac{B}{z-3}$$

we obtain

$$f(z) = \frac{-\frac{1}{3}}{z} + \frac{\frac{1}{3}}{z-3}$$

For this part we need to express $f(z)$ as a series in powers of $1/z$. We note that the first term above already has this form, so we move on to the second term, $\frac{1}{3}(z-3)^{-1}$:

$$\frac{\frac{1}{3}}{z-3} = \frac{\frac{1}{3}}{-3+z} = \frac{1}{-3} \cdot \frac{\frac{1}{3}}{1-\frac{z}{3}} = -\frac{1}{9}\left(1 + \frac{z}{3} + \frac{z^2}{9} + \frac{z^3}{27} + \cdots\right)$$

The sum of the two expressions then gives us the desired Laurent series:

$$f(z) = -\frac{1}{3}z^{-1} - \frac{1}{9} - \frac{1}{27}z - \frac{1}{81}z^2 - \frac{1}{243}z^3 - \cdots - (\tfrac{1}{3})^{n+2}z^n + \cdots = -\sum_{n=-1}^{\infty} \frac{z^n}{3^{n+2}}$$

The region of validity is all z for which $\frac{1}{3}|z| < 1$ (or $|z| < 3$) and $z \neq 0$. That is, inside the circle $|z| = 3$, $z \neq 3$. This region is a virtual annular ring $0 < |z| < 3$ (Figure 8.5a).

(b) $z = 3$: In this case we are looking for powers of $(z-3)$, so we write

$$\frac{-\frac{1}{3}}{z} + \frac{\frac{1}{3}}{z-3} = \frac{-\frac{1}{3}}{3+z-3} + \frac{\frac{1}{3}}{z-3} = -\frac{1}{9} \cdot \frac{1}{1+\dfrac{z-3}{3}} + \frac{\frac{1}{3}}{z-3}$$

which is then expanded into an infinite series using the geometric series

$$-\frac{1}{9}\left(1 - \frac{z-3}{3} + \frac{(z-3)^2}{9} - \cdots\right) + \frac{\frac{1}{3}}{z-3}$$

Therefore,

$$f(z) = \tfrac{1}{3}(z-3)^{-1} - \tfrac{1}{9} + \tfrac{1}{27}(z-3) - \tfrac{1}{81}(z-3)^2 + \tfrac{1}{243}(z-3)^3 - \cdots - (-\tfrac{1}{3})^{n+2}(z-3)^n + \cdots$$

The region of validity is the virtual annular ring $0 < |z-3| < 3$. (See Figure 8.5b.)

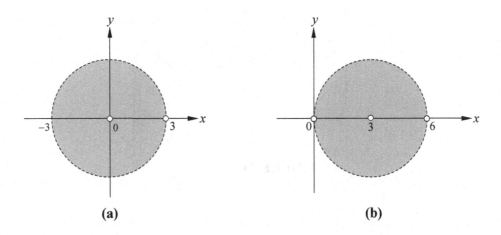

(a)   (b)

FIGURE 8.5

EXAMPLE 4

Consider the function $g(z) = \dfrac{z^2 - 4z}{(z-2)(z-3)^2}$. Use partial fractions and algebra to obtain functions in terms of $z-3$, then expand in Laurent series about $z = 3$. Obtain the region of convergence. Use this form to determine the region of validity for this series representation for $g(z)$.

SOLUTION
The correct set-up for this problem is

$$g(z) = \frac{A}{z-2} + \frac{B}{z-3} + \frac{C}{(z-3)^2} = \frac{z^2 - 4z}{(z-2)(z-3)^2}.$$

or

$$A(z-3)^2 + B(z-2)(z-3) + C(z-2) \equiv z^2 - 4z$$

By substituting the values 2, 3, then 0 for z, and solving for A, B, and C one obtains $A = -4$, $B = 5$, and $C = -3$. Thus

$$g(z) = \frac{-4}{z-2} + \frac{5}{z-3} + \frac{-3}{(z-3)^2} = \frac{-4}{(z-3)+1} + \frac{5}{z-3} + \frac{-3}{(z-3)^2}$$

The Laurent series about $z = 3$ is then

$$g(z) = \frac{-3}{(z-3)^2} + \frac{5}{z-3} - 4 + 4(z-3) - 4(z-3)^2 + 4(z-3)^3 + \cdots + (-1)^{n+1}4(z-3)^n + \cdots$$

The series converges for $z \neq 3$ and $|z-3| < 1$, thus for z in the virtual annular ring $0 < |z-3| < 1$. (See Figure 8.6.)

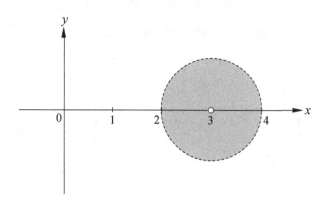

FIGURE 8.6

We close this section by proving Theorem 6.17 on removable singularities, postponed until now. Recall that it essentially states that if $f(z)$ is analytic at all points in a neighborhood N of c except at c, and if $\lim f(z) = d$ exists as $z \to c$, then the definition $f(c) = d$ makes $f(z)$ is *analytic at c and in N*. Thus we must prove that the derivative of $f(z)$ exists at $z = c$. By Laurent's Theorem, $f(z)$ has a Laurent series of the form

$$\cdots + \frac{b_m}{(z-c)^m} + \cdots + \frac{b_3}{(z-c)^3} + \frac{b_2}{(z-c)^2} + \frac{b_1}{z-c} + a_0 + a_1(z-c) + a_2(z-c)^2 + \cdots + a_n(z-c)^n + \cdots$$

valid in N for $z \neq c$. Now if any one of the coefficients b_m were nonzero, the limit of $f(z)$ would not exist as $z \to c$. Therefore, $b_1 = b_2 = b_3 = \cdots = b_m \cdots = 0$ and $f(z) = a_0 + a_1(z-c) + a_2(z-c)^2 + \cdots$. By hypothesis, the limit of $f(z)$ as $z \to c$ is d, so $a_0 = d$. Thus the derivative of $f(z)$ at $z = c$ is

$$\lim_{z \to c} \frac{f(z) - f(c)}{z - c} = \lim_{z \to c} \frac{a_1(z-c) + a_2(z-c)^2 + a_3(z-c)^3 \cdots}{z - c} = \lim_{z \to c} [a_1 + a_2(z-c) + a_3(z-c)^2 + \cdots] = a_1$$

Therefore, $f'(c)$ exists and f is analytic at c, and at all other points in N.

RESIDUE THEORY

In order to appreciate the significance of what follows, it is important for you to work through the following experiment (which should be a relatively easy exercise). You may have to review Cauchy's integral formula in order to complete it.

NUMERICAL EXPERIMENT

(1) Use Cauchy's integral formula to evaluate the integrals of

$$f(z) = \frac{1}{z(z-3)} \quad \text{and} \quad g(z) = \frac{z^2 - 4z}{(z-2)(z-3)^2}.$$

over C, which for $f(z)$ is the circle $|z| = 3$, then the circle $|z - 3| = 3$, and for $g(z)$, C is the circle $|z - 3| = 1$ (see Figures 8.4, 8.5, and 8.6). [**Hint:** For the first integral, the only singular point inside C is $z = 0$; in this case, write the integrand in the form $\dfrac{1/(z-3)}{z-0}$ and apply Cauchy's formula with $f(z) = 1/(z-3)$ and $z_0 = 0$. Use a similar technique at the singular point $z = 3$, and at the singular point $z = 3$ for $g(z)$ [here you will have to take a derivative and use Cauchy's formula for $n = 1$.] Place your results in Table 1 below (answer given for the first result).

(2) In each of the three cases, locate the coefficient a_{-1} of the Laurent series expansion of each function found in Examples 3 and 4, and complete Table 1. Did you notice anything?

	$f(z)$ for $z = 0$	$f(z)$ for $z = 3$	$g(z)$ for $z = 3$
Value of the integral over C	$2\pi i \cdot (-\frac{1}{3})$	$2\pi i \cdot (?)$	$2\pi i \cdot (?)$
Value of the coefficient a_{-1}			

TABLE 1

One of the important observations made by the pioneers in complex analysis was the fact alluded to in the numerical experiment above, namely that the term a_{-1} of the Laurent series for any function $f(z)$ yields the value for the integral of $f(z)$, often a difficult problem, and the curve of integration can be arbitrary. This leads to the important area of *contour integration* that enables one to evaluate (real) integrals that cannot be handled by standard methods. (This topic is covered in detail in the next chapter.)

To eliminate pathology, we consider only functions whose singular points are *isolated*. Prime examples are: $f(z) = 1/(z^n - 1)$ (n a positive integer), which has singularities at the nth roots of unity (points equally spaced on the unit circle), and $f(z) = 1/\sin z$, whose singularities are $z = n\pi$, $n = 0, \pm 1, \pm 2, \cdots$. A function that is not allowed would be one like $f(z) = 1/\sin(1/z)$ in a neighborhood of zero, whose singular points occur at $z = 0$ and at $z = 1/n\pi$, $n = \pm 1, \pm 2, \cdots$ (a sequence of points on the real axis converging to zero). Here, zero is not an isolated point of singularity.

Suppose, then, that $f(z)$ is analytic at all points in an open, simply-connected region R with boundary C, except at a set of interior isolated singular points $a, b, c, \cdots$ in R. In fact, the Bolzano-Weierstrass theorem implies that there can be only finitely many such points. (Do you see why?) Construct circles centered at each singular point inside R which exclude all the other points of singularity (Figure 8.7).

First, let's consider the case when R includes just one singular point a of $f(z)$ and let C_1 be a circle interior to an annular ring lying in the interior of R whose center is a (Figure 8.8). Thus $f(z)$ is analytic in that annular ring and has a Laurent series expansion about $z = a$ given by **(8.5)**

$$f(z) = \sum_{n=-\infty}^{\infty} a_n (z-a)^n \quad \text{where} \quad a_n = \frac{1}{2\pi i} \int_{C_1} \frac{f(z)}{(z-a)^{n+1}} dz \quad (n = 0, \pm 1, \pm 2, \pm 3, \cdots)$$

Set $n = -1$ in the formula for a_n to obtain

(8.6)
$$a_{-1} = \frac{1}{2\pi i} \int_{C_1} \frac{f(z)}{(z-a)^0} \, dz = \frac{1}{2\pi i} \int_{C_1} f(z) \, dz$$

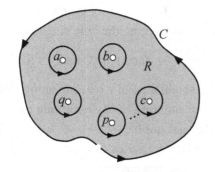

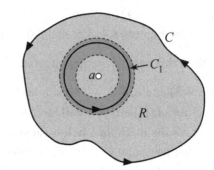

FIGURE 8.7   FIGURE 8.8

Now by Cauchy's theorem, the integral of $f(z)$ on the boundary C of R equals that on C_1. Thus the value for the integral of $f(z)$ on C is given by

(8.7)
$$\int_C f(z) \, dz = 2\pi i a_{-1}$$

Since the value a_{-1} alone determines the integral of $f(z)$ on C and thus all other components a_k of the integral of $f(z)$ are zero, it is called a **residue** of $f(z)$ at $z = a$ in R.

For the general case when R contains multiple singularities $a, b, c, \cdots$ of $f(z)$, as shown in Figure 8.7, then $f(z)$ also has a Laurent series expansion at each point $a, b, c, \cdots$, with the respective coefficients a_{-1}, $b_{-1}, c_{-1}, \cdots$ of the Laurent series expanded about those points. The following theorem is now evident.

THEOREM 8.8 (The Residue Theorem)
Suppose R is an open, simply-connected region, and that $f(z)$ is analytic in R except for a finite number of interior singular points $z = a, b, c, \cdots$. Then the integral of $f(z)$ on the boundary C of R is $2\pi i$ times the sum of the residues of $f(z)$ in R. That is,

(8.8)
$$\boxed{\int_C f(z) \, dz = 2\pi i (a_{-1} + b_{-1} + c_{-1} + \cdots)}$$

Proof: By Cauchy's theorem in the form **(5.6)** the integral of $f(z)$ over C is the sum of the integrals of $f(z)$ over each of the circles centered at each singular point. By **(8.7)** these integrals have the respective values $2\pi i a_{-1}, 2\pi i b_{-1}, 2\pi i c_{-1}, \cdots,$, and **(8.8)** follows. ⟍

EXAMPLE 5
  Show that
$$\int_0^{2\pi} \frac{d\theta}{(5 + 4\cos\theta)^3} = \frac{22\pi}{81}$$

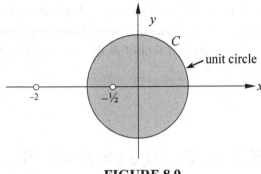

FIGURE 8.9

SOLUTION

We make a change of variable and use the formula for the integral over the unit circle $C : z = e^{i\theta}, 0 \leq \theta \leq \pi$ in reverse, to obtain the following (see Figure 8.9):

$$\cos \theta = \frac{e^{i\theta} + e^{-i\theta}}{2} = \frac{z + z^{-1}}{2} = \frac{z^2 + 1}{2z}, \qquad dz = ie^{i\theta} d\theta \quad \text{or} \quad d\theta = dz/iz$$

The given integral is then transformed into:

$$(*) \qquad I = \int_C \frac{1}{\left[5 + 4 \cdot \dfrac{z^2 + 1}{2z}\right]^3} \frac{dz}{iz} = \frac{1}{i} \int_C \frac{z^3}{\left[z\left(5 + \dfrac{4z^2 + 4}{2z}\right)\right]^3} \frac{dz}{z}$$

$$= \frac{1}{i} \int_C \frac{z^2 dz}{(5z + 2z^2 + 2)^3} = \frac{1}{i} \int_C \frac{z^2 dz}{(2z^2 + 5z + 2)^3}$$

In order to proceed we need to factor the denominator of the integrand and expand into partial fractions so that the Laurent series can be determined. We obtain the form

$$\frac{z^2}{(2z+1)^3(z+2)^3} = \frac{A}{2z+1} + \frac{B}{(2z+1)^2} + \frac{C}{(2z+1)^3} + \frac{D}{z+2} + \frac{E}{(z+2)^2} + \frac{F}{(z+2)^3}$$

Although this looks forbidding, there are simplifying features to be made before we actually solve for the constants A–F. The singularities occur at $z = -\frac{1}{2}, -2$. Since only the point $z = -\frac{1}{2}$ lies inside C (as shown in Figure 8.9), we need the residue at just that one point. We then seek the Laurent series in powers of $(z + \frac{1}{2})$. Note that by factoring out the "2" in the terms for A, B, and C, we already have the negative powers of $(z + \frac{1}{2})$ [we need only the term $(z + \frac{1}{2})^{-1}$]. The terms involving D, E, and F will not lead to negative powers of $(z + \frac{1}{2})$; together they contribute terms of the form

$$a_0 + a_1(z + \tfrac{1}{2}) + a_2(z + \tfrac{1}{2})^2 + a_3(z + \tfrac{1}{2})^3 + \cdots$$

So we do not need D, E, and F. It follows that, once we determine A, B, and C, the Laurent series of the integrand of I is

$$\frac{C/8}{(z + \frac{1}{2})^3} + \frac{B/4}{(z + \frac{1}{2})^2} + \frac{A/2}{z + \frac{1}{2}} + a_0 + a_1(z + \tfrac{1}{2}) + a_2(z + \tfrac{1}{2})^2 + a_3(z + \tfrac{1}{2})^3 + \cdots$$

Thus, the required residue is $A/2$, and it is only necessary to solve for A. This value is found to be $\frac{22}{81}$ (details left to the reader), and the residue is $\frac{11}{81}$. By **(8.8)**, the value of the integral is

$$I = \frac{1}{i} \cdot 2\pi i \cdot \frac{11}{81} = \frac{22\pi}{81}$$

NOTE: It should be pointed out that the method of residues to evaluate integrals is useful in a practical sense only for more difficult integrals, such as the one above. For less complicated functions, one can just as easily use Cauchy's integral formula to evaluate the integral instead of going to the trouble of expanding the function in Laurent series. The next section simplifies the process of finding residues, which then makes the method of residues easier to apply. ⟍

EVALUATING RESIDUES FOR QUOTIENTS OF ANALYTIC FUNCTIONS

Consider the problem of finding the residue of $\tan z$ at the singular point $z = \pi/2$. Since $\tan z$ is the quotient of the analytic functions $\sin z$ and $\cos z$, the direct procedure suggested by Taylor's theorem poses the problem of first expanding sine and cosine into powers of $z - \pi/2$, then dividing one power series by another to obtain the desired Laurent series. Certain rules do exist for such operations on series, but we have chosen to omit this topic. Fortunately, a different procedure is available.

In general, let's consider the quotient of two analytic functions $f(z)$ and $g(z)$ (both analytic at $z = c$), and suppose that, by Taylor's theorem the denominator can be expressed as

$$g(z) = \sum_{n=0}^{\infty} a_n (z - c)^n$$

Suppose, also, that the first k coefficients of the power series for $g(z)$ are zero, and that $a_k \neq 0$. That is,

$$g(z) = a_k(z - c)^k + a_{k+1}(z - c)^{k+1} + a_{k+2}(z - c)^{k+2} + \cdots$$

In this case the binomial $z - c$ is said to be a **factor** of $g(z)$ of **order** k, and we define c to be a **zero** of $g(z)$ **of order** k (called **simple** if $k = 1$).

THEOREM 8.9

Let $f(z)$ and $g(z)$ be analytic at $z = c$, and let $F(z) = f(z)/g(z)$. Then if c is a zero of $g(z)$ of order $k > 0$ and $f(c) \neq 0$, c is a pole of order k of the quotient $F(z)$.

Proof: By hypothesis, $g(z) = (z - c)^k G(z)$ where $G(z) = a_k + a_{k+1}(z - c) + \cdots$ and $a_k \neq 0$. Then $G(c) \neq 0$ and by continuity, $G(z) \neq 0$ in some neighborhood N of c. Thus, by the quotient rule, $f(z)/G(z)$ is differentiable in N and is therefore analytic at $z = c$. By Taylor's theorem, $f(z)/G(z)$ has a power series expansion of the form

$$\frac{f(z)}{G(z)} = b_0 + b_1(z - c) + b_2(z - c)^2 + \cdots$$

where $b_0 \neq 0$ (since $f(c) \neq 0$). Divide both sides by $(z - c)^k$ to obtain the Laurent series

$$\frac{f(z)}{(z - c)^k G(z)} = \frac{f(z)}{g(z)} = \frac{b_0}{(z - c)^k} + \frac{b_1}{(z - c)^{k-1}} + \cdots + \frac{b_{k-1}}{(z - c)} + b_k + b_{k+1}(z - c) + \cdots$$

and, by definition, c is a pole of order k of $f(z)/g(z)$. ⟍

Alternate formulas for residues as mentioned earlier can be obtained, which render Theorem 8.9 more useful. You can participate in establishing them by working Problem 16. They are as follows:

If $F(z)$ has a simple pole at $z = c$ then

(8.10)
$$a_{-1} = \lim_{z \to c} (z - c)F(z)$$

More generally, if $F(z)$ has a pole of order k at c then its residue at c is given by

(8.11)
$$a_{-1} = \frac{1}{(k-1)!} \lim_{z \to c} \left[(z - c)^k F(z)\right]^{(k-1)}$$

where $[F(z)]^{(n)}$ denotes the nth derivative of $F(z)$ if $n > 0$ (or $F(z)$ itself if $n = 0$).

EXAMPLE 6

Find the residue of $\tan z$ at $z = \pi/2$.

SOLUTION

In order to expand $\cos z$ about $z = \pi/2$, we use the series for $\sin z$ about $z = 0$ (which is valid for all z), as follows:

$$\cos z = -\sin(z - \frac{\pi}{2}) = -(z - \frac{\pi}{2}) + \frac{(z - \frac{\pi}{2})^3}{3!} - \frac{(z - \frac{\pi}{2})^5}{5!} + \cdots$$

Thus $\cos z$ has a zero of order 1 at $\pi/2$ with $\sin \pi/2 \neq 0$. By Theorem 8.9 $\tan z$ has a simple pole at $\pi/2$. By **(8.10)**,

$$a_{-1} = \lim_{z \to \pi/2} (z - \frac{\pi}{2}) \tan z = \lim_{z \to \pi/2} (z - \frac{\pi}{2}) \frac{\sin z}{\cos z} = \lim_{z \to \pi/2} \left(\frac{\sin z}{\frac{\cos z}{z - \pi/2}} \right) = -1$$

using L'Hospital's rule in the denominator.

EXAMPLE 7

Find the residues of the function $\dfrac{e^z}{z^3(z - 2)}$ at each of the two points of singularity:
(a) $z = 0$
(b) $z = 2$.

SOLUTION

(a) $z = 0$: By Theorem 8.9 this point is a pole of order 3, hence **(8.11)** becomes

$$\frac{1}{2!} \lim_{z \to 0} \left[z^3 \cdot \frac{e^z}{z^3(z - 2)} \right]^{(2)} = \frac{1}{2} \lim_{z \to 0} \left[\frac{e^z}{z - 2} \right]'' = \frac{1}{2} \lim_{z \to 0} \left[\frac{e^z(z - 3)}{(z - 2)^2} \right]'$$

$$= \frac{1}{2} \lim_{z \to 0} \left(\frac{e^z(z - 3)(z - 2)^2 + e^z(1)(z - 2)^2 - e^z(z - 3)(2)(z - 2)}{(z - 2)^4} \right)$$

$$= \frac{1}{2} \cdot \frac{e^0(-3)(-2)^2 + e^0(-2)^2 - e^0(-3)(2)(-2)}{(-2)^4}$$

$$= \frac{1}{2} \cdot \frac{-12 + 4 - 12}{16} = -\frac{5}{8}$$

(b) $z = 2$: This is a simple pole, so we use **(8.10)**:

$$\lim_{z \to 2} (z - 2) \cdot \frac{e^z}{z^3(z - 2)} = \lim_{z \to 2} \frac{e^z}{z^3} = \frac{e^2}{8}$$

EXAMPLE 8

Use **(8.11)** to find the residue of the function $\dfrac{z^3}{(2z+1)^3(z+2)^3}$ at the singularity $z = -\frac{1}{2}$. (The integral of this function was computed in the solution of Example 5, and the residue was found using partial fractions.)

SOLUTION

The function can be written as $\dfrac{z^3/8(z+2)^3}{(z+\frac{1}{2})^3}$, so the point $z = -\frac{1}{2}$ is a pole of order 3 (Theorem 8.9). Thus **(8.11)** becomes

$$a_{-1} = \frac{1}{2!}\lim_{z\to -1/2}\left[(z+\tfrac{1}{2})^3 \cdot \frac{z^2/8(z+2)^3}{(z+\tfrac{1}{2})^3}\right]^{(2)} = \frac{1}{16}\lim_{z\to -1/2}\left[\frac{z^2}{(z+2)^3}\right]''$$

$$= \frac{1}{16}\lim_{z\to -1/2}\left[\frac{2z(z+2)^3 - z^2 \cdot 3(z+2)^2}{(z+2)^6}\right]'$$

$$= \frac{1}{16}\lim_{z\to -1/2}\left[\frac{-z^2+4z}{(z+2)^4}\right]'$$

$$= \frac{1}{16}\lim_{z\to -1/2}\frac{2z^2-16z+8}{(z+2)^5}$$

$$= \frac{1}{16}\cdot\frac{\frac{1}{2}+8+8}{(\frac{3}{2})^5} = \frac{1}{16}\cdot\frac{\frac{33}{2}}{\frac{243}{32}} = \frac{11}{81}$$

EXAMPLE 9

Evaluate the integral $\displaystyle\int_0^{2\pi}\frac{d\theta}{(4+2\sin\theta)^2}\,d\theta$.

SOLUTION

As before, we transform this integral to a complex integral on the unit circle $C: z = e^{i\theta}$, $0 \le \theta \le 2\pi$.

$$\sin z = \frac{z - z^{-1}}{2i} = \frac{z^2-1}{2iz} \qquad \text{and} \qquad dz = ie^{i\theta}d\theta \ \ (\text{or } d\theta = \frac{dz}{iz})$$

By substitution (with I as the given integral),

$$I = \int_C \frac{1}{\left(4+2\cdot\dfrac{z^2-1}{2iz}\right)^2}\cdot\frac{dz}{iz} = \int_C \frac{(iz)^2}{(4iz+z^2-1)^2}\cdot\frac{dz}{iz} = -\frac{1}{i}\int_C \frac{z\,dz}{(z^2+4iz-1)^2}$$

We need to factor the denominator, so using the quadratic formula, the zeros are found to be $c_1 = (-2+\sqrt{3})i$ and $c_2 = (-2-\sqrt{3})i$, with c_1 the only singularity of the integrand that lies within C. Thus

$$I = -\frac{1}{i}\int_C \frac{z\,dz}{(z-c_1)^2(z-c_2)^2}$$

where the singularity c_1 is a pole of order 2. Using **(8.11)**,

$$a_{-1} = \frac{1}{1!}\lim_{z\to c_1}\left[(z-c_1)^2\cdot\frac{z}{(z-c_1)^2(z-c_2)^2}\right]' = \lim_{z\to c_1}\left[\frac{z}{(z-c_2)^2}\right]' = \lim_{z\to c_1}\frac{-z-c_2}{(z-c_2)^3} = \frac{-c_1-c_2}{(c_1-c_2)^3}$$

A little algebra reduces this number to $-\sqrt{3}/18$. Thus we obtain

$$I = -\frac{1}{i} \cdot 2\pi i \cdot \frac{-\sqrt{3}}{18} = \frac{\pi\sqrt{3}}{9}$$

A final example will show how to solve a type of problem (like some at the end of this chapter) in the most convenient way.

EXAMPLE 10

Show that if $f(z) = \dfrac{\cos z(\sin z + 1)}{z^3}$, the point $z = 0$ is a pole of order 3, and use **(8.11)** to compute the residue at 0.

SOLUTION

Since the numerator is nonzero at $z = 0$, by Theorem 8.9, it is a pole of order 3. Using **(8.11)** with $k = 3$, we have

$$a_{-1} = \frac{1}{2} \lim_{z \to 0} \left(\frac{z^3 \cos z(\sin z + 1)}{z^3} \right)^{(2)} = \frac{1}{2} \lim_{z \to 0} \left[(\cos z(\sin z + 1) \right]''$$

It saves some work here if we use the differentiation formula from calculus $(fg)'' = f''g + 2f'g' + fg''$:

$$a_{-1} = \frac{1}{2} \lim_{z \to 0} \left[(-\cos z(\sin z + 1) + 2(-\sin z)(\cos z) + (\cos z)(-\sin z) \right] = \frac{1}{2} \cdot (-1) = -\frac{1}{2}$$

BERNOULLI NUMBERS AND A LAURENT SERIES FOR cot z

An interesting discovery was made by a famous mathematician of the 18th century named John Bernoulli. The coefficients of the series expansion of $z/(e^z - 1)$ about $z = 0$ (which are called *Bernoulli numbers*) can be easily determined explicitly. We call the function $z/(e^z - 1)$ the *Bernoulli function*, denoted by $B(z)$.

First, note that

$$B(z) \equiv \frac{z}{e^z - 1} = \frac{1}{1 + \dfrac{z}{2!} + \dfrac{z^2}{3!} + \dfrac{z^3}{4!} + \cdots}$$

Let $g(z)$ denote the infinite series in the denominator. Note that $g(z)$ is analytic and nonzero at $z = 0$, so the Bernoulli function is analytic at $z = 0$ and has a Taylor series representation $\sum a_n z^n$, where $a_n = B^{(n)}(0)/n!$ for $n = 0, 1, 2, \cdots$. Since $B(z) = 1/g(z)$, then $a_0 = 1$, and $a_1 = B'(0) = -g'(0)/[g(0)]^2 = -\frac{1}{2}$. Also,

$$a_2 = B''(0)/2! = -\frac{1}{2} \cdot \left\{ g''(0) \cdot [g(0)]^2 - g'(0) \cdot 2g(0)g'(0) \right\} / [g(0)]^4 = \frac{1}{12}.$$

In Problem 20 you are asked to show that $z/(e^z - 1) + z/2 \equiv (z/2)\coth(z/2)$, which is an even function. Hence $(z/2)\coth(z/2) = 1 + a_2 z^2 + a_3 z^3 + \cdots$ has only even powers. Thus $a_3 = a_5 = a_7 \cdots = 0$. This proves the basic form for the series expansion for $B(z)$:

(8.12) $$\frac{z}{e^z - 1} = \frac{1}{1 + \dfrac{z}{2!} + \dfrac{z^2}{3!} + \dfrac{z^3}{4!} + \cdots} = 1 - \frac{z}{2} + \frac{B_1}{2!}z^2 + \frac{B_2}{4!}z^4 + \frac{B_3}{6!}z^6 + \cdots \qquad (B_1 = \frac{1}{6})$$

where B_n ($n = 1, 2, 3, \cdots$) are real numbers. These values are the **Bernoulli numbers**.

An interesting result is that the Bernoulli numbers satisfy a relatively simple identity. It can be shown that for all $n \geq 2$

(8.13) $$(2n + 1)B_n = \frac{2n-1}{2} - \binom{2n+1}{2}B_1 - \binom{2n+1}{4}B_2 - \binom{2n+1}{6}B_3 - \cdots - \binom{2n+1}{2n-2}B_{n-1}$$

This is a *recursive identity* valid for $n \geq 2$, and it generates the Bernoulli numbers B_n for $n \geq 2$, starting with $B_1 = \frac{1}{6}$. For example, if $n = 2$ we obtain

$$5B_2 = \frac{3}{2} - \binom{5}{2}B_1 = \frac{3}{2} - \frac{5 \cdot 4}{1 \cdot 2}B_1 = \frac{3}{2} - \frac{10}{6} = -\frac{1}{6} \quad \Rightarrow \quad B_2 = -\frac{1}{30}$$

NOTE: The identity (8.13) can be established by using the Cauchy product method for the product of two infinite series. From (8.12) we obtain

$$\left(1 + \frac{z}{2!} + \frac{z^2}{3!} + \frac{z^3}{4!} + \frac{z^4}{5!} + \cdots + \frac{z^n}{n!} + \cdots\right) \cdot \left(1 + a_1 z + a_2 z^2 + a_3 z^3 + a_4 z^4 + \cdots + a_n z^n + \cdots\right) \equiv 1$$

where $a_1 = -\frac{1}{2}$, $a_{2n} = \dfrac{B_n}{(2n)!}$, and $a_{2n+1} = 0$, ($n = 1, 2, 3, \cdots$). Setting the coefficients of z^n in the product equal to zero (for $n \geq 1$), we obtain a form that is equivalent to (8.13). ⬳

One can obtain from (8.12) an explicit series for several trigonometric functions in terms of Bernoulli numbers, such as $\tan z$, $\sec z$, and $\cot z$. Although the formula for B_n is not in closed form, it does provide a mechanism for numerically computing as many terms of each of these series as we desire. We illustrate with $\cot z$.

Start with the series for $\coth z$, from which $\cot z$ can be determined:

$$\coth z = \frac{e^z + e^{-z}}{e^z - e^{-z}} = \frac{(e^z + e^{-z})e^z}{(e^z - e^{-z})e^z} = \frac{e^{2z} + 1}{e^{2z} - 1}$$

$$= \frac{(e^{2z} - 1) + 2}{e^{2z} - 1} = 1 + \frac{2}{e^{2z} - 1} = 1 + \frac{1}{z} \cdot \frac{2z}{e^{2z} - 1}$$

$$= 1 + \frac{1}{z}\left(1 - \frac{2z}{2} + \sum_{n=1}^{\infty} \frac{B_n(2z)^{2n}}{(2n)!}\right)$$

This reduces to,

(8.14) $$\coth z = \frac{1}{z} + \sum_{n=1}^{\infty} \frac{B_n(2z)^{2n}}{(2n)! \, z}$$

Since $\cot z = \dfrac{\cos z}{\sin z} = \dfrac{e^{iz} + e^{-iz}}{(e^{iz} - e^{-iz})/i} = i \coth iz$, we obtain from (8.14)

$$\cot z = i\left(\frac{1}{iz} + \sum_{n=1}^{\infty} \frac{B_n(2iz)^{2n}}{(2n)! \, iz}\right)$$

or

(8.15) $$\cot z = \frac{1}{z} + \sum_{n=1}^{\infty} \frac{(-1)^n B_n(2z)^{2n}}{(2n)! \, z} = \frac{1}{z} - \frac{2^2 B_1}{2!}z + \frac{2^4 B_2}{4!}z^3 - \frac{2^6 B_3}{6!}z^5 + \cdots$$

The final example will demonstrate once again the power of complex variables, where we make use of (8.15). We prove the classical result $\sum \dfrac{1}{n^2} = \dfrac{\pi^2}{6}$ due to Euler.

EXAMPLE 11

Show that for any positive integer k, if C_k is the circle $|z| = p_k$, where $p_k = (2k+1)\dfrac{\pi}{2}$, the integral in k

$\displaystyle\int_{C_k} \frac{\cot z\, dz}{z^2}$ converges to zero as $k \to \infty$. Find the residues of $\cot z / z^2$ inside C_k, and evaluate $\sum 1/n^2$.

SOLUTION

The first part requires proving that $|\cot z|$ is bounded for points on C_k ($k = 1, 2, 3, \cdots$). With $z = x + iy$, observe that

$$|\cot z| = \frac{|\cos z|}{|\sin z|} = \frac{|\cos x \cosh x - i \sin y \sinh y|}{|\sin x \cosh y + i \cos y \sinh y|} = \frac{\sqrt{\cos^2 x \cosh^2 x + \sin^2 y \sinh^2 y}}{\sqrt{\sin^2 x \cosh^2 x + \cos^2 y \sinh^2 y}} = \sqrt{\frac{\cos^2 x + \sinh^2 y}{\sin^2 x + \sinh^2 y}}$$

(where the identity $\cosh^2 y - \sinh^2 y = 1$ was used). Thus,

$$|\cot z|^2 - 1 = \frac{\cos^2 x + \sinh^2 y}{\sin^2 x + \sinh^2 y} - 1 = \frac{\cos 2x}{\sin^2 x + \sinh^2 y}$$

If $|\cot z|$ is *not* bounded for $z \in C_k$, we can find a sequence $\{z_m\}$ lying on some C_k such that

(∗)
$$\frac{\cos 2x_m}{\sin^2 x_m + \sinh^2 y_m} \to \infty$$

But $\cos 2x_m$ is bounded, so we conclude that $\sin^2 x_m + \sinh^2 y_m \to 0$. Hence, both $\sin^2 x_m$ and $\sinh^2 y_m$ converge to 0. Thus $y_m \to 0$. If a subsequence of $\{z_m\}$ lies on the same circle C_k for some k, then we can assume that $\{z_m\}$ converges to a point $z_0 = x_0 + iy_0$ on that circle, hence $y_0 = 0$ and $x_0 = \pm(2k+1)\pi/2$. Thus $\sin^2 x_0 = \sin^2(k\pi + \pi/2) = 1$, contradicting $\sin^2 x_m \to \sin^2 x_0 = 0$. Then it follows that we can find a member z_k of $\{z_m\}$ lying on some circle C_k, with $k \to \infty$. In this case, for each such k we must have $x_k^2 + y_k^2 = p_k^2$ and $x_k = \pm(p_k^2 - y_k^2)^{1/2}$. Consider the definition $u_k = (p_k^2 - y_k^2)^{1/2} - p_k$. By algebra,

$$u_k = \frac{(p_k^2 - y_k^2)^{1/2} - p_k}{1} \cdot \frac{(p_k^2 - y_k^2)^{1/2} + p_k}{(p_k^2 - y_k^2)^{1/2} + p_k} = \frac{p_k^2 - y_k^2 - p_k^2}{(p_k^2 - y_k^2)^{1/2} + p_k} = \frac{-y_k^2}{(p_k^2 - y_k^2)^{1/2} + p_k}$$

Since $\left|(p_k^2 - y_k^2)^{1/2} + p_k\right| \geq |p_k| = |2k+1|\pi/2 > 1$ we obtain $|u_k| < y_k^2$ forcing $u_k \to 0$ as $k \to \infty$. Thus:

$$\sin^2 x_k = \sin^2(p_k^2 - y_k^2)^{1/2} = \sin^2(p_k + u_k)$$

$$= \sin^2[(2k+1)\pi/2 + u_k]$$

$$= \sin^2[k\pi + \pi/2 + u_k]$$

$$= \sin^2(\pi/2 + u_k) \to 1$$

again contradicting $\sin^2 x_k \to 0$. We can then conclude that for some real number B, $|\cot z| < B$ holds for all k and all points $z \in C_k$. It now follows that the integrals mentioned converge to zero as $k \to \infty$:

$$\left|\int_{C_k} \frac{\cot z\, dz}{z^2}\right| \leq \int_{C_k} \frac{|\cot z||dz|}{|z|^2} \leq \frac{B}{p_k^2} l(C_k) = \frac{B \cdot 2\pi p_k}{p_k^2} = \frac{2\pi B}{p_k} = \frac{4B}{2k+1}$$

establishing the desired result. Finally, we evaluate the residues of $\cot z / z^2$. The singularities of

$$\frac{\cot z}{z^2} = \frac{\cos z}{z^2 \sin z}$$ occur at $z = 0, \pm\pi, \pm 2\pi, \pm 3\pi, \cdots$, which, except for $z = 0$, are simple poles. Thus

$$\operatorname*{Res}_{z=n\pi} \frac{\cot z}{z^2} = \lim_{z \to n\pi} \frac{(z - n\pi)\cos z}{z^2 \sin z} = \frac{1}{n^2 \pi^2} \lim_{z \to n\pi} \frac{\cos z}{\dfrac{\sin z}{z - n\pi}} = \frac{1}{n^2 \pi^2} \lim_{z \to n\pi} \frac{\cos z}{\dfrac{\cos z}{1}} = \frac{1}{n^2 \pi^2}$$

The residue at $z = 0$ will be evaluated directly from the series **(8.15)**:

$$\frac{\cot z}{z^2} = \frac{1}{z^2}\left(\frac{1}{z} - \frac{2^2 B_1}{2!} z + \frac{2^4 B_2}{4!} z^3 - \cdots\right) = \frac{1}{z^3} - \frac{1}{3z} + \frac{2^4 B_2}{4!} z - \cdots$$

and we find the residue to be $-\frac{1}{3}$. The singularities of $\cot z / z^2$ inside the circle C_k are $0, \pm\pi, \pm 2\pi$, $\pm 3\pi, \cdots, \pm m\pi$ where $m < k\pi/2 < m + 1$, so by the residue theorem

$$\int_{C_k} \frac{\cot z\, dz}{z^2} = 2\pi i \left(-\frac{1}{3} + \frac{1}{\pi^2} + \frac{1}{2^2 \pi^2} + \frac{1}{3^2 \pi^2} + \cdots + \frac{1}{m^2 \pi^2}\right)$$

$$+ \left(\frac{1}{(-1)^2 \pi^2} + \frac{1}{(-2)^2 \pi^2} + \frac{1}{(-3)^2 \pi^2} + \cdots + \frac{1}{(-m)^2 \pi^2}\right)$$

and we obtain

$$\left|-\frac{1}{3} + 2\sum_{n=1}^{m} \frac{1}{n^2 \pi^2}\right| = \left|\frac{1}{2\pi i}\int_{C_k} \frac{\cot z\, dz}{z^2}\right|$$

As $k \to \infty$ (and therefore $m \to \infty$),

$$-\frac{1}{3} + 2\sum_{n=1}^{\infty}\frac{1}{n^2 \pi^2} = 0 \qquad \text{or} \qquad \sum_{n=1}^{\infty}\frac{1}{n^2} = \frac{\pi^2}{6}$$

PROBLEMS

1. Find the function in closed form to which the following series converges $(0 < |z| < 1)$:

$$\frac{3}{z^2} - \frac{4}{z} + 1 - z + z^2 - z^3 + \cdots$$

2. Find the function in closed form to which the following series converges $(0 < |z| < 1)$:

$$\frac{1}{3z^2} - \frac{2}{z} + 1 - z + z^2 - z^3 + \cdots$$

3. Find the function in closed form to which the following series converges $(0 < |z + 1| < 1)$:

$$\frac{1}{(z+1)^2} + (z+1)^2 + (z+1)^6 + (z+1)^{10} + \cdots$$

4. Expand the function $\dfrac{1}{z(z+1)}$ in Laurent series about $z = 0$ that is valid for $0 < |z + 1| < 1$.

5. Expand the function $\dfrac{\ln z}{(z-1)^4}$ in Laurent series about $z = 1$ valid for $0 < |z - 1| < 1$ and show that this point is a pole of order 3.

6. Using the Laurent series already computed, find the residue of the function of
   (a) Problem 1 at $z = 0$
   (b) Problem 5 at $z = 1$.

7. Find the residue of the function of
   (a) Problem 1 at $z = 0$ using **(8.11)** with $k = 2$.
   (b) Problem 5 at $z = 1$ using **(8.11)** with $k = 3$.

8. Find the residue of the function of Problem 2 at $z = 0$ using the Laurent series expansion.

9. Find the residue of the function of Problem 2 at $z = 0$ using **(8.11)**.

10. (a) Find the poles of the function $f(z) = \dfrac{z+2}{z^6 - z^4}$ and find their orders.

(b) Find the residues of $f(z)$. [**Hint:** For $z = 0$, use the formula $g^{(3)}(0)h(0) + 3g''(0)h'(0) + 3g'(0)h''(0) + h^{(3)}(0)$ for the third derivative of $g(z)h(z)$ evaluated at $z = 0$, where $g(z) = z + 2$ and $h(z) = (z^2 - 1)^{-1}$.]

(c) If C is the circle $|z| = \frac{1}{2}$, find $\displaystyle\int_C \dfrac{z+2}{z^6 - z^4}\, dz$.

11. Show that $\displaystyle\int_0^{2\pi} \dfrac{d\theta}{10 + 6\sin\theta} = \dfrac{\pi}{4}$.

12. Show that $\displaystyle\int_0^{2\pi} \dfrac{d\theta}{3 + \sin\theta - 2\cos\theta} = \pi$.

13. Evaluate $\displaystyle\int_0^{2\pi} \dfrac{d\theta}{13 + 12\sin\theta}$.

14. Evaluate $\displaystyle\int_0^{2\pi} \dfrac{2\cos\theta}{5 + 4\cos\theta}\, d\theta$.

15. Evaluate $\displaystyle\int_0^{2\pi} \dfrac{d\theta}{(5 - 3\sin\theta)^2}\, d\theta$.

16. Prove **(8.11)** by assuming that the Laurent series expansion at c is given by

$$F(z) = \frac{a_{-k}}{(z-c)^k} + \frac{a_{-k+1}}{(z-c)^{k-1}} + \cdots + \frac{a_{-2}}{(z-c)^2} + \frac{a_{-1}}{z-c} + a_0 + a_1(z-c) + a_2(z-c)^2 + \cdots$$

Multiply both sides by $(z - c)^k$ and using termwise differentiation to obtain the $(k-1)^{\text{st}}$ derivative of the result, evaluated at $z = c$.

17. Expand the following function in a Laurent series about the singularity $z = 0$ and find the residue:

$$f(z) = \frac{z - \sin z}{z^6}$$

18. Use **(8.11)** to find the residue of Problem 17 at $z = 0$.

19. Find the residue of $f(z) = \csc z/z^2$ at $z = 0$ given that 0 is a pole of order 3.

20. Derive the identity $(z/2)\coth(z/2) \equiv z/(e^z - 1) + z/2$. [**Hint:** Start with $\tanh z = \sinh z/\cosh z = (e^z - e^{-z})/(e^z + e^{-z})$.]

21. Compute B_3 and B_4.

22. A result often found in handbooks or in calculus textbooks is

$$\cot z = \frac{1}{z} - \frac{z}{3} - \frac{z^3}{45} - \frac{2z^5}{945} - \frac{z^7}{4725} - \cdots$$

which gives the first five terms of the series for $\cot z$ explicitly. Verify this result using **(8.15)** and the result from Problem 21.

23. Using the identity $\csc z = \cot z/2 - \cot z$, establish the identiy $\csc z = \dfrac{1}{z} + \displaystyle\sum_{n=1}^{\infty} \dfrac{(-1)^n(2 - 4^n)B_n z^{2n-1}}{(2n)!} = \dfrac{1}{z} + \dfrac{z}{6} + \dfrac{7z^3}{360} + \cdots$ Then check this result against that appearing in handbooks or on line.

SPECIAL TOPICS

IN COMPLEX VARIABLES

We have seen how complex variable theory possesses special powers that go beyond ordinary real variable calculus. The theory developed by the early pioneers was found to solve many problems in applied mathematics by the researchers who followed. Some applications include important areas in engineering mathematics and industry. Liquid-flow dynamics, weather forecasting, electrical engineering, and aerodynamics are just a few examples. Since we do not intend to cover many such applications in this book, particularly those that involve advanced applied mathematics, this chapter will be limited to concepts that should be accessible to beginning students. Other sources should be consulted for a more comprehensive treatment.

IMPROPER INTEGRALS: A QUICK REVIEW

One area that demonstrates the power of complex variables involves the evaluation of certain improper integrals. For example, consider the integral

$$\int_0^\infty e^{-x}dx$$

It is improper since the upper limit is infinite. It's definition (as with all improper integrals) is based on limits. In this case one defines

$$\int_0^\infty e^{-x}dx = \lim_{r\to\infty}\int_0^r e^{-x}dx$$

The integral can then be evaluated by ordinary finite integration and the evaluation of a well-known limit:

$$\int_0^\infty e^{-x}dx = \lim_{r\to\infty}\int_0^r e^{-x}dx = \lim_{r\to\infty}\left(-e^{-x}\Big|_0^r\right) = \lim_{r\to\infty}\left(-e^{-r}+e^{-0}\right) = \lim_{r\to\infty}\left(1-e^{-r}\right) = 1$$

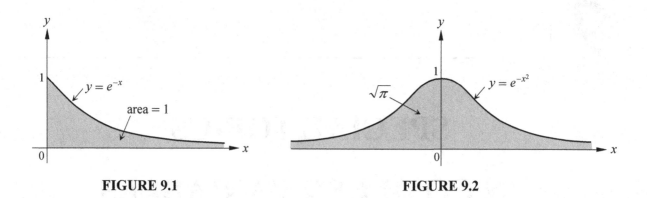

FIGURE 9.1   **FIGURE 9.2**

This number represents the area of the infinite region under the curve $y = e^{-x}$, $x \geq 0$ (Figure 9.1).

Sometimes integrals are mathematically difficult because the anti-derivatives cannot be determined in closed form, even with advanced methods. Such integrals must be evaluated individually using special techniques, and in many cases only numerical solutions are possible. (In fact, exact answers can be determined for only a relatively few integrals.) As an example, we mention one important integral which plays a key role in statistics, whose exact value is well known. It gives the area under the *normal curve*, defined classically as the curve $y = e^{-x^2}$. It takes a special argument (which we do not give here) to establish the result, as illustrated in Figure 9.2,

$$\int_{-\infty}^{\infty} e^{-x^2} dx = \sqrt{\pi}$$

As you may recall from your calculus course, improper integrals involving both upper and lower limits, as in the above example, must be split at some convenient point to form two integrals, one for each of the two improper limits. To illustrate, consider

$$\int_{-\infty}^{\infty} \frac{e^{-x}}{(e^{-x}+1)^2} dx = \int_{-\infty}^{0} \frac{e^{-x}}{(e^{-x}+1)^2} dx + \int_{0}^{\infty} \frac{e^{-x}}{(e^{-x}+1)^2} dx$$

In this case, an anti-derivative of the integrand equals $(e^{-x} + 1)^{-1}$ and we obtain the result

$$\lim_{r \to -\infty} \frac{1}{e^{-x}+1}\Big|_{r}^{0} + \lim_{r \to \infty} \frac{1}{e^{-x}+1}\Big|_{0}^{r} = \lim_{r \to -\infty}\left(\frac{1}{2} - \frac{1}{e^{-r}+1}\right) + \lim_{r \to \infty}\left(\frac{1}{e^{-r}+1} - \frac{1}{2}\right) = \left(\frac{1}{2} - 0\right) + \left(\frac{1}{0+1} - \frac{1}{2}\right) = 1$$

Much use will be made of another fact concerning improper integrals. If the integrand is an odd function $[f(-x) = -f(x)$ for all $x]$, then the integral under the curve from $-\infty$ to 0 is the negative of the integral from 0 to ∞, and

$$\int_{-\infty}^{\infty} f(x)dx = 0$$

If the function is even $[f(-x) = f(x)]$, then due to symmetry the integrals on $(-\infty, 0]$ and $[0, \infty)$ are equal, so it follows that

$$\int_{-\infty}^{\infty} f(x)dx = 2\int_{0}^{\infty} f(x)dx$$

Thus, in the case of the normal curve,

$$\int_{0}^{\infty} e^{-x^2} dx = \frac{\sqrt{\pi}}{2}$$

CONTOUR INTEGRATION

One method for obtaining exact values for a variety of improper integrals involves Cauchy's theorem and residue theory. The basic idea is to select some closed curve C_r (as shown in Figure 9.3) that includes the real axis from $-r$ to $+r$, to find by residue theory the complex integral on C_r that corresponds to the real improper integral we are trying to evaluate, and then let $r \to \infty$. If one can show that the integral over the non-real part of C_r approaches zero as $r \to \infty$, then the remaining part is a complex integral over the real interval $[-r , r]$ which, as $r \to \infty$, converges to the value of the real integral we seek. This process is indicated by the flow-chart appearing in the insert of Figure 9.3, a procedure called **contour integration**. Usually, the easy part is finding the residues and evaluating the complex integral over the closed curve C_r, and the hard part is showing that the integral over the non-real part of C_r approaches zero.

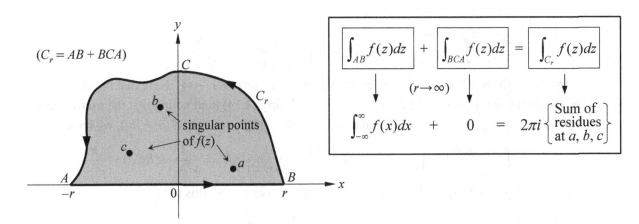

$(C_r = AB + BCA)$

FIGURE 9.3

To show how this works, let's evaluate a standard improper integral you no doubt had in calculus by using the established method of limits (as in the preceding section), and then, after that, by using this new method of contour integration. Suppose we consider the integral

$$\int_0^\infty \frac{dx}{x^2 + 1}$$

An anti-derivative of $(1 + x^2)$ is $\tan^{-1} x$, so we obtain the one-line solution (no real need to use contour integration here):

$$\int_0^\infty \frac{dx}{x^2 + 1} = \lim_{r \to \infty} \int_0^r \frac{dx}{x^2 + 1} = \lim_{r \to \infty} \tan^{-1} x \Big|_0^r = \lim_{r \to \infty} (\tan^{-1} r - \tan^{-1} 0) = \frac{\pi}{2}$$

Now let's see if we get the same result using contour integration. By following the schematic diagram in Figure 9.3, we first consider the complex integral

$$\int_{C_r} \frac{dz}{z^2 + 1}$$

We choose the curve of integration C_r to be the semicircle as shown in Figure 9.4, consisting of the base AB and semicircular arc BCA centered at the origin. The singular points of the integrand are the zeroes of $z^2 + 1$, or $z = \pm i$, of which, $z = i$ is the only one inside C_r for $r > 1$. By the residue theorem,

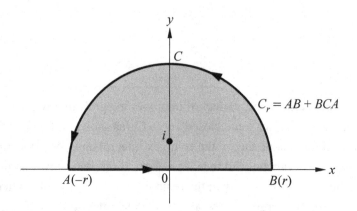

FIGURE 9.4

$$\int_C \frac{dz}{z^2+1} = 2\pi i \cdot \operatorname{Res}\{(z^2+1)^{-1} \text{ at } z=i\} = 2\pi i \cdot \lim_{z \to i}\left[(z-i)\cdot\frac{1}{(z+i)(z-i)}\right] = 2\pi i \cdot \frac{1}{2i} = \pi$$

We obtain

(∗) $$\int_{AB}^r \frac{dz}{z^2+1} + \int_{BCA} \frac{dz}{z^2+1} = \pi \quad \text{or} \quad \int_{-r}^r \frac{dx}{x^2+1} + \int_{BCA} \frac{dz}{z^2+1} = \pi$$

It remains to show that the integral over BCA approaches zero as $r \to \infty$. Note that the curve BCA has the parametric form $z = re^{it}$, $0 \le t \le \pi$. Thus, $|z| = r$. Using **(4.14)** (and noting that the triangle inequality implies for complex numbers a and b that $|a+b| \ge ||a|-|b|| = |a|-|b|$ if $|a| \ge |b|$), we have

$$\left|\int_{BCA} \frac{dz}{z^2+1}\right| \le \int_{BCA} \left|\frac{1}{z^2+1}\right| |dz| \le \int_{BCA} \frac{1}{|z|^2-1} |dz| \le \frac{1}{r^2-1} l(BCA) = \frac{\pi r}{r^2-1}$$

Since the limit of the expression on the right equals zero as $r \to \infty$, this proves

$$\lim_{r \to \infty} \int_{BCA} \frac{dz}{z^2+1} = 0$$

Thus, taking the limit as $r \to \infty$ in (∗),

$$\int_{-\infty}^\infty \frac{dx}{x^2+1} + 0 = \pi$$

or, by symmetry,

$$\int_0^\infty \frac{dx}{x^2+1} = \frac{\pi}{2}$$

Some problems in this area have been designed for you to work on at the end of this chapter. A few more examples will be presented for further guidance.

EXAMPLE 1

Evaluate the integral $\int_0^\infty \frac{dx}{x^4+1}$.

SOLUTION

The same semicircle as used before will work here, namely $C_r = AB + BCA$ (Figure 9.4). Here the singular points of $(z^4+1)^{-1}$ all lie on the unit circle, spaced at equal distances apart: $e^{\pi i/4}$, $e^{3\pi i/4}$, $e^{5\pi i/4}$, and $e^{7\pi i/4}$, as shown in Figure 9.5 ($r > 1$). The only ones inside C_r are $e^{\pi i/4} \equiv c_1$ and $e^{3\pi i/4} \equiv c_2$, and they are both simple poles, so the residues at these two points (by L'Hospital's rule) are as follows:

$$\operatorname{Res}\{(z^4+1)^{-1} \text{ at } z=c_1\} = \lim_{z \to c_1} \frac{z-c_1}{z^4+1} = \lim_{z \to c_1} \frac{1}{4z^3} = \frac{1}{4e^{3\pi i/4}} = \frac{1}{4(-\sqrt{2}/2 + \sqrt{2}/2\, i)} = \frac{-1-i}{4\sqrt{2}}$$

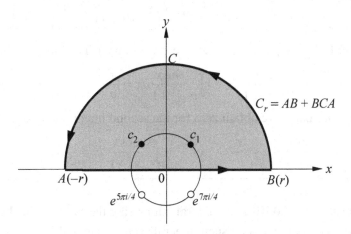

FIGURE 9.5

$$\text{Res } \{(z^4+1)^{-1} \text{ at } z=c_2\} = \lim_{z \to c_2} \frac{z-c_2}{z^4+1} = \lim_{z \to c_2} \frac{1}{4z^3} = \frac{1}{4e^{9\pi i/4}} = \frac{1}{4(\sqrt{2}/2 + \sqrt{2}/2\, i)} = \frac{1-i}{4\sqrt{2}}$$

Hence

$$2\pi i\left(\frac{-1-i}{4\sqrt{2}} + \frac{1-i}{4\sqrt{2}}\right) = \frac{2\pi i\,(-2i)}{4\sqrt{2}} = \frac{\pi}{\sqrt{2}} \qquad \text{and} \qquad \int_{C_r} \frac{dz}{z^4+1} = \frac{\pi}{\sqrt{2}}$$

Thus

(*) $$\int_{AB} \frac{dz}{z^4+1} + \int_{BCA} \frac{dz}{z^4+1} = \frac{\pi}{\sqrt{2}}$$

As before,

$$\left|\int_{BCA} \frac{dz}{z^4+1}\right| \le \int_{BCA} \left|\frac{1}{z^4+1}\right| |\,dz\,| \le \frac{l(BCA)}{r^4-1} = \frac{\pi r}{r^4-1} \to 0 \quad \text{as} \quad r \to \infty$$

Therefore, taking the limit in (*) as $r \to \infty$,

$$\int_{-\infty}^{\infty} \frac{dx}{x^4+1} = \frac{\pi}{\sqrt{2}} \qquad \text{or} \qquad \int_0^{\infty} \frac{dx}{x^4+1} = \frac{\pi}{2\sqrt{2}} = \frac{\pi\sqrt{2}}{4}$$

EXAMPLE 2

Evaluate the integral $\displaystyle\int_0^{\infty} \frac{\cos x}{x^2+1}\,dx$.

SOLUTION

We use the same semicircle C_r as in Figure 9.5. Since $\cos z = \frac{1}{2}(e^{iz} + e^{-iz})$, we first evaluate the integral of $e^{iz}/(z^2+1)$ over C_r. The singular points of $e^{iz}/(z^2+1)$ are $z = \pm i$, of which only $z = i$ lies inside C_r. Since i is a simple pole,

$$\int_{C_r} \frac{e^{iz}}{z^2+1}\,dz = 2\pi i \cdot \operatorname*{Res}_{at\, z=i} \frac{e^{iz}}{z^2+1} = 2\pi i \cdot \lim_{z \to i} \frac{(z-i)e^{iz}}{z^2+1} = 2\pi i \cdot \lim_{z \to i} \frac{e^{iz}}{z+i} = 2\pi i \cdot \frac{e^{-1}}{2i} = \pi e^{-1}$$

Again, we show that the limit of the integral over the curve BCA as $r \to \infty$ is zero, where BCA has the parametric form $z = re^{ti}$, $0 \le t \le \pi$.

$$\left|\int_{BCA} \frac{e^{iz}}{z^2+1}\right| \le \int_{BCA} \left|\frac{1}{z^2+1}\right| \left|e^{iz}\right| |dz| \le \int_{BCA} \frac{1}{|z|^2-1}\, e^{-y}\, |dz| \le \frac{1}{r^2-1} \cdot l(BCA)$$

(since $y \ge 0$). Thus

$$\left|\int_{BCA} \frac{e^{iz}}{z^2+1}\right| \le \frac{\pi r}{r^2-1} \to 0 \qquad \text{(as } r \to \infty)$$

and

$$\int_{AB}\frac{e^{iz}}{z^2+1}dz + \int_{BCA}\frac{e^{iz}}{z^2+1}dz = \frac{\pi}{e} \quad \text{or} \quad \int_{-r}^{r}\frac{\cos x}{x^2+1}dx + i\int_{-r}^{r}\frac{\sin x}{x^2+1}dx + \int_{BCA}\frac{e^{iz}}{z^2+1}dz = \frac{\pi}{e}$$

(This was obtained from Euler's equation $e^{iz} = e^{ix} = \cos x + i\sin x$.) The limit as $r\to\infty$ produces

$$\int_{-\infty}^{\infty}\frac{\cos x}{x^2+1}dx + i\int_{-\infty}^{\infty}\frac{\sin x}{x^2+1}dx = \frac{\pi}{e}$$

Equating real and imaginary parts, we obtain zero for the second integral above, and, for the first,

$$\int_{-\infty}^{\infty}\frac{\cos x}{x^2+1}dx = \frac{\pi}{e} \quad \text{or} \quad \int_{0}^{\infty}\frac{\cos x}{x^2+1}dx = \frac{\pi}{2e}$$

A final example will show how to work an integral with a singular point on the boundary of the semi-circle C_r. It should be pointed out that not all such integrals can be handled in this manner, and obviously other curves besides semicircles are often used. A variety of methods have evolved which we do not cover here.

EXAMPLE 3

Show that $\int_{0}^{\infty}\frac{\sin x}{x}dx = \frac{\pi}{2}$.

SOLUTION

As in Example 2, since $\sin z = (e^{iz} - e^{-iz})/2i$, we first consider the integral of e^{iz}/z over the appropriate curve. Since the path of integration must avoid singular points, we try the compound curve $C_r = AB + BCD + DE + EFA$ as shown in Figure 9.6, made up of two line segments on the x-axis and two semi-circles centered at 0, one of radius ε (where later, $\varepsilon\to 0$), and the other of radius r (where $r\to\infty$).

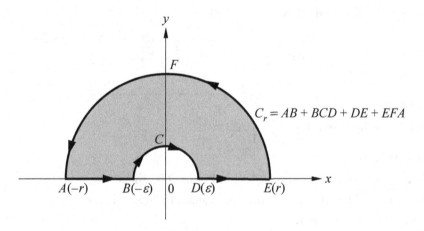

FIGURE 9.6

The point $z = 0$ is the only singular point of e^{iz}/z, and C_r does not pass through it. In fact, e^{iz}/z is analytic on and inside C_r. Hence

$$\int_{C_r}\frac{e^{iz}}{z}dz = 0 = \int_{AB}\frac{e^{iz}}{z}dz + \int_{BCD}\frac{e^{iz}}{z}dz + \int_{DE}\frac{e^{iz}}{z}dz + \int_{EFA}\frac{e^{iz}}{z}dz$$

Let's evaluate the integrals over BCD first. With $BCD = -DCB$ and $z = \varepsilon e^{ti}$, $0\le t\le\pi$,

$$\int_{BCD}\frac{e^{iz}}{z}dz = -\int_{0}^{\pi}\frac{e^{i(\varepsilon\cos t + \varepsilon i\sin t)}}{\varepsilon e^{it}}\cdot\varepsilon i e^{it}dt = -\int_{0}^{\pi}ie^{-\varepsilon\sin t}[\cos(\varepsilon\cos t) + i\sin(\varepsilon\cos t)]dt$$

The limit as $\varepsilon \to 0$ produces

$$\lim_{\varepsilon \to 0} \int_{BCD} \frac{e^{iz}}{z} dz = -i \int_0^\pi e^0 (\cos 0 + i \sin 0) dt$$

$$= -i \int_0^\pi dt = -\pi i$$

For the integral over EFA, which we want to show approaches zero as $r \to \infty$, we obtain (with $z = re^{it}$, $0 \le t \le \pi$):

$$\left| \int_{EFA} \frac{e^{iz}}{z} dz \right| \le \int_{EFA} \frac{|e^{iz}|}{|z|} |dz| = \int_0^\pi \frac{e^{-y}}{r} \left| \frac{dz}{dt} \right| dt = \int_0^\pi \frac{e^{-r \sin t}}{r} \left| rie^{it} \right| dt = \int_0^\pi e^{-r \sin t} dt$$

By change of variable ($t' = \pi - t$), one can show that $\int_{\pi/2}^\pi e^{-r \sin t} dt = \int_0^{\pi/2} e^{-r \sin t} dt$. Hence

$$\left| \int_{EFA} \frac{e^{iz}}{z} dz \right| \le 2 \int_0^{\pi/2} e^{-r \sin t} dt$$

Furthermore, by observation, the graph of $y = \sin t$ for $0 \le t \le \pi/2$ lies above the line segment joining $(0, 0)$ and $(\pi/2, 1)$, and one concludes that $\sin t \ge 2t/\pi$. Thus if $a = 2/\pi$, we have $-\sin t \le -at$, and

$$\left| \int_{EFA} \frac{e^{iz}}{z} dz \right| \le 2 \int_0^{\pi/2} e^{-rat} dt = 2 \left(-\frac{1}{ar} e^{-rat} \Big|_0^{\pi/2} \right) = \frac{2}{ar} (1 - e^{-r}) < \frac{2}{ar}$$

with zero limit as $r \to \infty$.

Thus since

$$\int_{-r}^\varepsilon \frac{e^{ix}}{x} dx + \int_\varepsilon^r \frac{e^{ix}}{x} dx + \int_{BCD} \frac{e^{iz}}{z} dz + \int_{EFA} \frac{e^{iz}}{z} dz = 0$$

then as $r \to \infty$ and $\varepsilon \to 0$, one obtains

$$\int_{-\infty}^0 \frac{e^{ix}}{x} dx + \int_0^\infty \frac{e^{ix}}{x} dx - \pi i + 0 = 0 \qquad \text{or} \qquad \int_{-\infty}^\infty \frac{e^{ix}}{x} dx = \pi i$$

Therefore,

$$\int_{-\infty}^\infty \frac{\cos x}{x} dx + i \int_{-\infty}^\infty \frac{\sin x}{x} dx = \pi i$$

Since $\sin x/x$ is an even function, equating imaginary parts produces

$$\int_0^\infty \frac{\sin x}{x} dx = \frac{\pi}{2}$$

FUNCTIONS AND MAPPINGS: REVIEW

Two areas of mathematics you may have studied are *linear algebra* and *transformation geometry*. In linear algebra you considered functions (or mappings) from one vector space to another ($f: V \to U$, where V and U are vector spaces). This can also be written in matrix form as $u = f(v) \equiv Av$ where v denotes the given vector $\langle x_1, x_2, \cdots, x_n \rangle$ in V, u is the *image* vector $\langle y_1, y_2, \cdots, y_n \rangle$ of v in U, and A is an $n \times n$ matrix $[a_{ij}]$. For example, in two dimensions, this becomes

(9.1)

$$\begin{cases} y_1 = a_{11} x_1 + a_{12} x_2 \\ y_2 = a_{21} x_1 + a_{21} x_2 \end{cases}$$

As we vary the constants a_{ij}, we obtain all possible linear (line-preserving) mappings in the plane that leave the origin fixed. Thus for any such mapping f, one has $f: \mathbb{R}^2 \to \mathbb{R}^2$ where $\mathbb{R}^2 = \mathbb{R} \times \mathbb{R}$. Thus f is a mapping from one coordinate plane to another. In geometry, the two planes are regarded as identical, and f is thought of as a method for transforming objects in the plane in some way.

Consider a couple numerical examples, where, for convenience we replace x_1 and x_2 by x and y, and y_1 and y_2 by u and v:

$$f: \begin{cases} u = x + 3y \\ v = x - 4y \end{cases} \qquad \text{and} \qquad g: \begin{cases} u = 4x - 3y \\ v = -8x + 6y \end{cases}$$

Both these mappings map lines into lines. For example, consider the image of the line $y = 2x - 1$ under f. By substitution into the equations for f we find:

$$u = x + 3(2x - 1) = 7x - 3$$
$$v = x - 4(2x - 1) = -7x + 4$$

In order to put this into $v = mu + b$ form, we must eliminate the parameter x. We solve the first equation for x and substitute into the second:

$$x = (u + 3)/7 \quad \Rightarrow \quad v = -7(u + 3)/7 + 4 \quad \Rightarrow \quad v = -u + 1$$

Thus, f maps the line $y = 2x - 1$ (slope 2) to the line $v = -u + 1$ (slope -1).

The details for finding the image of this same line under g are as follows:

$$u = 4x - 3(2x - 1) = -2x + 3, \ v = -8x + 6(2x - 1) = 4x - 6 \Rightarrow x = (u - 3)/(-2) \Rightarrow v = 4(u - 3)/(-2) - 6$$

or $v = -2u$. Therefore, g maps the line $y = 2x - 1$ to the line $v = -2u$. What is peculiar about g is that *every* line maps to $v = -2u$. Can you verify this? Thus, g maps the entire xy-plane into a single line in the uv-plane. In this case, g is clearly not one-to-one, and it has no inverse. On the other hand, the inverse of f does exist and is given by

$$x = \tfrac{4}{7}u + \tfrac{3}{7}v$$
$$y = \tfrac{1}{7}u - \tfrac{1}{7}v$$

This may be found by solving the system for x and y in terms of u and v using elimination or Cramer's rule. If you have a knowledge of matrices, note that, alternatively,

$$\begin{vmatrix} 1 & 3 \\ 1 & -4 \end{vmatrix}^{-1} = -\frac{1}{7} \begin{vmatrix} -4 & -3 \\ -1 & 1 \end{vmatrix} = \begin{bmatrix} \tfrac{4}{7} & \tfrac{3}{7} \\ \tfrac{1}{7} & -\tfrac{1}{7} \end{bmatrix}$$

The distinction between the two mappings can be seen by simply looking at the *determinant* of the matrix of coefficients of the two maps (which by definition equals $a_{11}a_{22} - a_{12}a_{21}$ in general). The one for f is nonzero, which characterizes all one-to-one linear functions. The one for g is zero (a degenerate map).

Some linear mappings are particularly useful in geometry. We are going to follow the geometric tradition of referring to a one-to-one mapping of a plane to itself (or portion thereof) as a **transformation**. Also, a mapping that preserves angle measure is called **conformal**, and one that preserves distance, **isometric**.

In general, a *conformal linear transformation* is always given by one the two forms

(9.2) $\qquad \begin{cases} u = ax - by \\ v = bx + ay \end{cases} \qquad \text{or} \qquad \begin{cases} u = ax + by \\ v = bx - ay \end{cases} \qquad (a^2 + b^2 \neq 0)$

For example, the mapping

$$\begin{cases} u = 3x - 4y \\ v = 4x + 3y \end{cases}$$

takes any two perpendicular lines in the xy-plane and maps them into perpendicular lines in the uv-plane (see Problem 11). If, in addition, $a^2 + b^2 = 1$, the mapping is isometric. In general, however, either of the mappings in (9.2) rotates a figure and either shrinks or enlarges it. Thus, it is called a **dilation-rotation**.

EXAMPLE 4

The rotation transformation often encountered in calculus has the form

$$\begin{cases} x' = x\cos\theta - y\sin\theta \\ y' = x\sin\theta + y\cos\theta \end{cases}$$

Interpret this mapping in terms of complex variables by showing it has the form $z' = cz$ where $c = a + bi$, is a fixed complex number having magnitude 1, and show that this mapping is distance-preserving.

SOLUTION

Let $c = e^{i\theta} = \cos\theta + i\sin\theta$. Then by the rules of multiplication for complex numbers, with $z = x + iy$ and $z' = x' + iy'$, $cz = (\cos\theta + i\sin\theta)\cdot(x + iy) = \cos\theta\cdot x - \sin\theta\cdot y + i(\sin\theta\cdot x + \cos\theta\cdot y)$. Equating real and imaginary parts, the mapping given above results. Also, $|c| = \sqrt{\cos^2\theta + \sin^2\theta} = 1$, which agrees with the condition $a^2 + b^2 = 1$ for an isometric mapping.

One important type of transformation in geometry is nonlinear. Under this transformation, some lines are mapped into circles instead of lines. Known as a **circular inversion**, it is given by

(9.3) $F: u = \dfrac{r^2 x}{x^2 + y^2}, \quad v = \dfrac{r^2 y}{x^2 + y^2}$ or $F(x, y) = \left(\dfrac{r^2 x}{x^2 + y^2}, \dfrac{r^2 y}{x^2 + y^2} \right)$

(for real $r \neq 0$ and $(x, y) \neq (0, 0)$. It will be discussed later in connection with complex variables. An inversion is defined for all points (x, y) except the origin. It maps the deleted xy-plane one-to-one onto the deleted uv-plane. By solving (9.3) for x and y in terms of u and v, we find that the inverse transformation has *exactly the same form* as the original mapping:

(9.4) $F^{-1}: x = \dfrac{r^2 u}{u^2 + v^2}, \quad y = \dfrac{r^2 v}{u^2 + v^2}$ or $F^{-1}(u, v) = \left(\dfrac{r^2 u}{u^2 + v^2}, \dfrac{r^2 v}{u^2 + v^2} \right)$

One can see that if the mapping F were applied twice, the result would be the same point (x, y) we started with {that is, $F[F(x, y)] \equiv F^2(x, y) = (x, y)$}. Thus $F^{-1} = F$ and F is its own inverse. Such a mapping is called an **involution**. (A more familiar example of an involution is a line reflection, as in a reflection across the x-axis; such a mapping is clearly its own inverse.)

To make things less abstract, let's work with (9.3) when $r = 2$. Thus

$$F: u = \frac{4x}{x^2 + y^2}, \quad v = \frac{4y}{x^2 + y^2} \qquad \text{or} \qquad F(x, y) = \left(\frac{4x}{x^2 + y^2}, \frac{4y}{x^2 + y^2} \right)$$

Suppose we want to find the image of the line having equation $y = -2x + 8$ (that is, $v = -2u + 8$ in terms of u and v), as illustrated in Figure 9.7. By substitution, $F[v = -2u + 8]$ becomes

$$\left(\frac{4y}{x^2 + y^2} \right) = -2 \left(\frac{4x}{x^2 + y^2} \right) + 8 \quad \Rightarrow \quad 4y = -8x + 8(x^2 + y^2) \quad \Rightarrow \quad x^2 + y^2 - x - \tfrac{1}{2}y = 0$$

We find that the line $y = -2x + 8$ maps to the *circle* $x^2 + y^2 - x - \tfrac{1}{2}y = 0$. Using standard algebra techniques, the equation of this circle can be put in the form

$$(x - \tfrac{1}{2})^2 + (y - \tfrac{1}{4})^2 = \tfrac{5}{16}$$

This represents a circle passing through $O(0, 0)$ having center $(\tfrac{1}{2}, \tfrac{1}{4})$ and radius $\sqrt{5}/4$, as shown in the figure. Its center lies on the line $y = \tfrac{1}{2}x$, which is perpendicular to the given line. This is true in general: a line not passing through the origin maps to a circle passing through O whose center C lies on the pependicular from O to the given line. Conversely, a circle passing through O with center C maps to a line perpendicular to the line of centers OC.

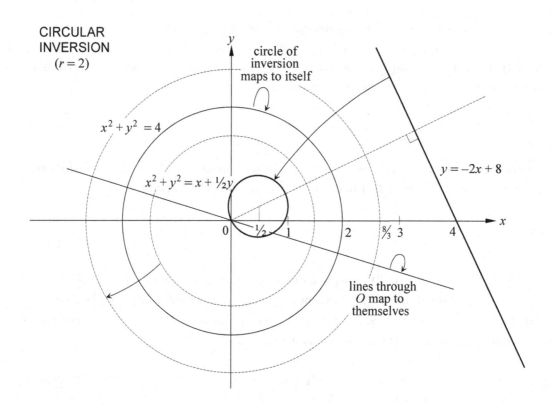

FIGURE 9.7

Other features of circular inversion are: The circle $x^2 + y^2 = r^2$ (called the **circle of inversion**) maps to itself, where each point is a **fixed point** (a point that is unchanged by the mapping). A circle centered at the origin inside the circle of inversion maps to a circle centered at the origin outside the circle of inversion (you can easily verify this by considering $x^2 + y^2 = s^2$, $s < r$). Also, a line passing through the origin maps to itself (but individual points are changed). Figure 9.7 shows these features for the case $r = 2$.

COMPLEX FUNCTIONS AS MAPPINGS

A function on the complex numbers $f: \mathbb{C} \to \mathbb{C}$ can be interpreted as a transformation of the complex plane. Thus if $w = f(z)$, then $w \equiv u + iv$ is the image of $z \equiv x + iy$. That is,

$$w = f(z) \quad \Rightarrow \quad u + iv = u(x, y) + iv(x, y) \qquad \text{or} \qquad \begin{cases} u = u(x, y) \\ v = v(x, y) \end{cases}$$

For example, if $w = 2z$ then $u + iv = 2(x + iy)$, and

$$\begin{cases} u = 2x \\ v = 2y \end{cases}$$

This example is a *dilation*, which doubles all distances. Unfortunately, the mappings induced by the elementary functions in complex variables, such as $\sin z$ and $\ln z$, are considerably more complicated. Let's begin with a few simpler functions, and gradually work our way up to more difficult situations.

Suppose we start with a map that just consists of multiplying a constant complex number c times the variable z, $w = cz$. This defines the **constant multiple** map. Thus if $c = a + bi$, then

$$cz = (a + bi)(x + iy) = ax - by + i(bx + ay)$$

and we obtain the mapping

$$\begin{cases} u = ax - by \\ v = bx + ay \end{cases}$$

Does this look familiar? You should recognize this as a *dilation-rotation* **(9.2)**. The effect of this mapping is the uniform *rotation* of all figures in the xy-plane, and the uniform shrinking or enlarging by the amount $|c|$ (a *dilation*). All this is in accordance with the fundamental property of complex multiplication discussed in Chapter 1: the magnitude of the product of two complex numbers equals the product of the magnitudes, and the argument of a product equals the sum of the arguments.

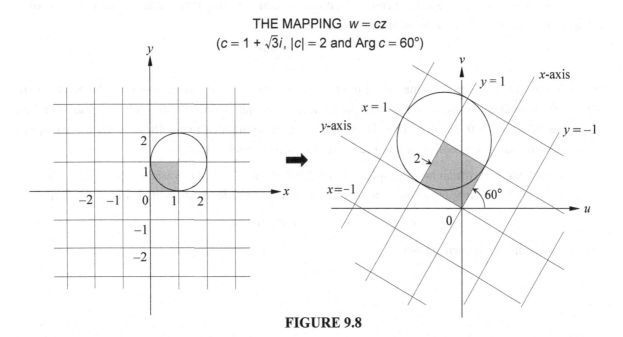

THE MAPPING $w = cz$
($c = 1 + \sqrt{3}i$, $|c| = 2$ and Arg $c = 60°$)

FIGURE 9.8

A specific example of this when $c = 1 + \sqrt{3}\,i$ is shown in Figure 9.8), where Arg $c = 60°$ and $|c| = 2$. Here,

$$\begin{cases} u = x - \sqrt{3}y \\ v = \sqrt{3}x + y \end{cases}$$

In this case, the mapping rotates all figures in the xy-plane through an angle of $60°$ and doubles their size.

An even simpler type of mapping is a **translation**, given by $w = z + z_0$. In terms of coordinates, $w = x + iy + x_0 + iy_0 = (x + x_0) + i(y + y_0)$; this mapping takes each point (x, y) in the direction (x_0, y_0) to the point $(x + x_0, y + y_0)$. Such mappings have the effect of moving every point in the complex plane in the exact same direction by the same exact distance. Intuitively, a translation may be described as a mapping that slides an object from one point to another, without rotation. Translations are clearly distance- and angle-preserving. If we combine this with a constant-multiple map, the following algebraic form results:

$$w = cz + d$$

Since this mapping has the same geometric properties as the constant-multiple map (a dilation-rotation), it maps lines to lines and circles to circles. It will be called a **complex linear** map (not to be confused with a linear mapping in general, which can map circles to ellipses).

The next mapping will show a surprisingly simple way to represent the circular inversion map in geometry. It is simply

$$w = \frac{r^2}{z}$$

where $r > 0$. To see this, observe that $\dfrac{r^2}{z} = \dfrac{r^2}{x + iy} = \dfrac{r^2(x - iy)}{x^2 + y^2} = \dfrac{r^2 x}{x^2 + y^2} + i \cdot \dfrac{-r^2 y}{x^2 + y^2}$

Thus

$$\begin{cases} u = \dfrac{r^2 x}{x^2 + y^2} \\ v = \dfrac{-r^2 y}{x^2 + y^2} \end{cases}$$

But for the minus sign, this is identical to **(9.3)**. The minus sign has the effect of mapping a point to its reflection in the x-axis. Thus the above mapping is a *geometric inversion followed by a reflection in the x-axis*. It will be called a **complex inversion**, to distinguish it from an ordinary circular inversion.

These two mappings are special cases of a more general mapping that can be defined in terms of simple algebra. It is the **fractional linear** mapping, given by:

$$w = \frac{az + b}{cz + d}$$

If $c \neq 0$, this equation defines a map from the deleted plane ($z \neq -d/c$) one-to-one onto the deleted plane ($z \neq a/c$). It need not be linear since lines can sometimes be mapped to circles, as in the case of an inversion (the case $a = d = 0$, $b > 0$, and $c = 1$). Can a line ever be mapped to an ellipse under these mappings? This and other questions will be left to the reader (see Problems 14, 15.)

Another mapping which has a simple algebraic definition in complex variables is $w = z^2$. Although quite elementary, this mapping has some unusual qualities. In order to study its features, observe that:

$$w = u + iv = (x + iy)^2 = x^2 - y^2 + 2xyi$$

Thus we obtain the **squaring map**

$$\begin{cases} u = x^2 - y^2 \\ v = 2xy \end{cases}$$

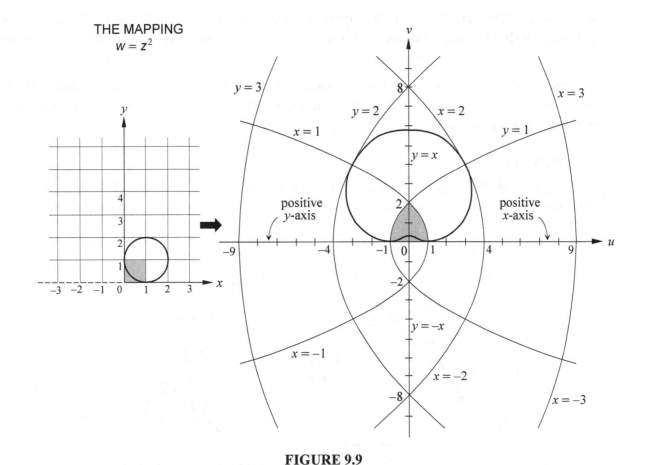

FIGURE 9.9

As its algebraic form suggests, the squaring map is not linear. For example, consider the vertical line $x = 2$. By substitution

$$u = 4 - y^2$$
$$v = 4y$$

where y varies over the real numbers (acting as a parameter). In order to eliminate y, we can obtain

$$y = v/4 \quad \Rightarrow \quad u = 4 - v^2/16$$

This is a *parabola* in the uv-plane, with horizontal axis. Its vertex is at $(4, 0)$, opening to the left (Figure 9.9). We soon find that all vertical lines map to parabolas with vertices on the u-axis, opening to the left. All horizontal lines ($y = a$, constant) also map to parabolas with vertices on the u-axis, but opening to the right, as shown in Figure 9.9. Thus we find that the positive x-axis maps to the positive u-axis, the positive y-axis maps to the negative u-axis, and the coordinate grid in the xy-plane maps to two systems of parabolas with vertices on the u-axis and opening in opposite directions. The circle $(x - 1)^2 + (y - 1)^2 = 1$ maps to an *indented oval*, resembling a curve often studied in calculus, the *limaçon*. (See in this connection Problem 17.)

Clearly, like its real counterpart, this function is not one-to-one. In order to achieve one-to-oneness, we must restrict the domain of the mapping in some way. As in real variables, such a region is not unique, but it is chosen to be as convenient and simple as possible. [Recall how the inverse sine function was defined in algebra.] In the case of $w = z^2$, we will restrict the complex number z to lie in the *upper half-plane* ($y > 0$), and to include numbers lying on the *non-negative real axis* ($x \geq 0$, $y = 0$), as indicated in Figure 9.9 (left diagram). This region is transformed one-to-one onto the entire uv-plane (see Problem 16).

One can observe that the two systems of parabolas intersect at right angles. They constitute a family of curves known as **orthogonal trajectories**. This property is no accident; it is shared by all mappings defined by analytic functions.

THEOREM 9.5

Suppose $f(z)$ is analytic at z_0 and that $f'(z_0) \neq 0$. Then the mapping $w = f(z)$ is conformal throughout some neighborhood of z_0.

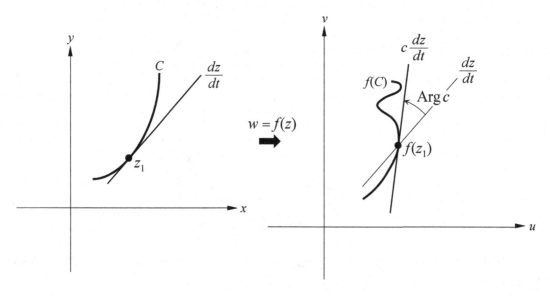

FIGURE 9.10

The proof is a simple application of the chain rule for complex variables and some geometry concerning tangents to parametric curves. Recall that if a smooth curve C in the xy-plane is given parametrically by $x = x(t)$, $y = y(t)$ for $a \leq t \leq b$ where $x(t)$ and $y(t)$ are (real) differentiable functions, then the derivatives dx/dt and dy/dt (if not both zero) give the components of the *tangent vector* to C. [That is, the directed line segment from $O(0, 0)$ to the point (dx/dt, dy/dt) is parallel to the tangent line to C at the point ($x(t)$, $y(t)$) .] The complex variable version of this is that the equation $z(t) \equiv x(t) + iy(t)$ represents C and that dz/dt is a complex number whose real and imaginary parts are the components of the tangent vector at $z(t)$, (See Figure 9.10.)

Regarding the chain rule used, it is the special one mentioned (and proved) in Chapter 3, directly following the general chain rule.

Proof of Theorem 9.5
Since $f'(z_0) \neq 0$ and f is analytic at z_0, $f'(z) \neq 0$ for all z in some neighborhood N of z_0. We consider first the effect the mapping $w = f(z)$ has on the tangent vector to a curve $z = z(t)$ in N. The image of a curve C in the xy-plane is given by $w(t) = f[z(t)]$ [denoted by $f(C)$]. Now differentiate $w(t)$ to find its tangent vector. We obtain by the chain rule

$$\frac{dw}{dt} = f'[z(t)]\frac{dz}{dt}$$

At a specific point in N lying on the curve z_1 [$= z(t_1)$], this becomes

$$\frac{dw}{dt} = f'(z_1)\frac{dz}{dt} = c\frac{dz}{dt}$$

where $c = f'(z_1) \neq 0$. The angle which the vector dz/dt makes with the x-axis is increased (or decreased) by the amount Arg c. If another curve C^* [$z = z^*(t)$] intersects C at z_1, the angle its tangent vector makes with the x-axis would be increased by exactly the same amount (Arg c), hence the angle between the two image curves $f(C)$ and $f(C^*)$ would be the same as that between the given curves C and C^* at $f(z_1)$. ⬚

EXAMPLE 5

Theorem 9.5 shows that the squaring map is angle-preserving (since $f(z) = z^2$ is analytic). But due to the multiplicative property for the square of a complex number, the argument of z^2 is twice the argument of z, seemingly a contradiction. Discuss.

SOLUTION

The theorem requires $f'(z_1) \neq 0$ in order to guarantee the conformal property for curves intersecting at z_1. But $f'(z) = 2z$, which is zero at $z = 0$, so there is no contradiction. At all other points, f is conformal.

The last mapping to be considered here is the **sine mapping**, the mapping induced by the sine function $w = \sin z$. By the definition given in Chapter 2, this mapping has the form

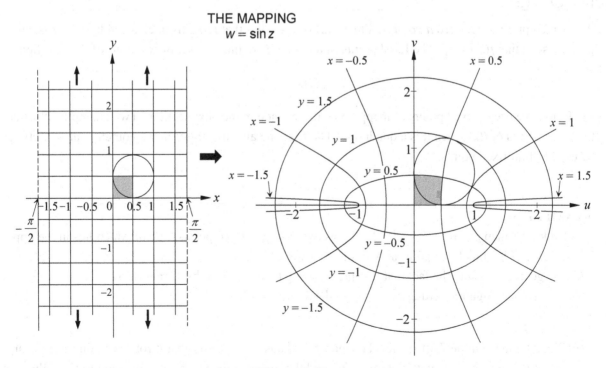

THE MAPPING
$w = \sin z$

Figure 9.11

$$\begin{cases} u = \sin x \cosh y \\ v = \cos x \sinh y \end{cases}$$

It maps the coordinate grid into a system of ellipses and hyperbolas as shown in Figure 9.11. [The figure shows the restricted domain of the mapping (the vertical strips $-\pi/2 < x \le \pi/2, y \ge 0$ and $-\pi/2 \le x < \pi/2, y < 0$) on which the sine map is one-to-one, as explored in Problem 20.] Since $\sin z$ is analytic, the mapping is conformal due to Theorem **9.5**. It maps all vertical lines to hyperbolas and horizontal lines to ellipses, and the hyperbolas are each orthogonal to the ellipses. These conics are also *confocal* (have the same foci). As an example, consider any vertical line $x = p$. By substitution

$$\begin{cases} u = \sin p \cosh y \\ v = \cos p \sinh y \end{cases}$$

By squaring and summing, these equations lead to

$$\frac{u^2}{\sin^2 p} - \frac{v^2}{\cos^2 p} = \cosh^2 y - \sinh^2 y = 1$$

which represents a hyperbola centered at $(0, 0)$, and having foci at the points $(\pm c, 0)$, where the formula for the foci of a hyperbola from elementary algebra produces

$$c = \sqrt{a^2 + b^2} = \sqrt{\sin^2 p + \cos^2 p} = 1$$

A similar analysis can be given for the images of the horizontal lines $y = q$, which are ellipses (Problem 19).

FIXED POINTS

As defined previously, a *fixed point* of a map $f(z)$ is any point that remains unchanged by f; that is, any point z_0 such that $f(z_0) = z_0$. The fixed points of any map f are thus the set of solutions of the equation

$$f(z) = z$$

Extensive use of fixed points of mappings will be made in the next section. Two examples illustrate the fixed points of $f(z)$ when it is a quadratic. The first one also illustrates how geometry can sometimes be used to find fixed points.

EXAMPLE 6

Consider the map given by $w = z^2 + 1$. It has two non-real fixed points, one of which lies in the upper half-plane. Find this fixed point in two different ways:

(a) by geometry, using the fact that it is the squaring map followed by a translation,

(b) by ordinary algebra, setting $z^2 + 1 = z$ and solving for z.

SOLUTION

(a) Let z lie in the upper half-plane. The map $z \rightarrow z^2$ has the property that it rotates a point z by its argument θ to the point z^2, with argument 2θ, and that result is then translated one unit to the right to obtain $z^2 + 1$ (see Figure 9.12, left diagram). If z is a fixed point, then $z = z^2 + 1$ and a translation of one unit to the right sends z^2 to $z^2 + 1$, or z, and we obtain the parallelogram $OPQR$ shown in the figure, where $OP = 1 = RQ$. Also, we must have equal angles at O and Q. Thus triangle ROQ is isosceles and $RO = RQ = 1$, or $r^2 = 1$, and $r = 1 = OQ$. Hence triangle OQR is equilateral and $\theta = 60°$ with Q (or z) equal to $\frac{1}{2} + \frac{1}{2}\sqrt{3}i$.

(b) Solving the quadratic equation $z^2 - z + 1 = 0$ using the quadratic formula, we find the two fixed points:

$$z = \frac{1 \pm \sqrt{1^2 - 4}}{2} = \frac{1}{2} \pm \frac{1}{2}\sqrt{3}i$$

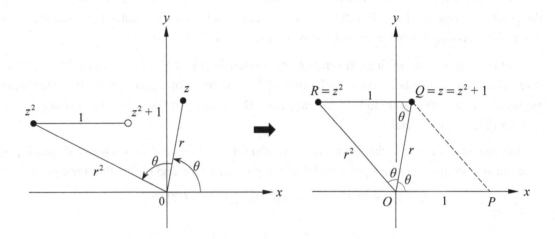

FIGURE 9.12

EXAMPLE 7

Find the fixed points of the function $g(z) = f(f(z))$ where $f(z) = z^2 + 1$.

SOLUTION

We need to solve the equation $g(z) = z$ for z. By algebra,

$$g(z) = (f(z))^2 + 1 = (z^2 + 1)^2 + 1 = z^4 + 2z^2 + 2$$

so we must solve the 4th degree equation $z^4 + 2z^2 + 2 = z$. An observation will save some work: note that if c is a fixed point of f, it is also a fixed point of g (since $g(c) = f(f(c)) = f(c) = c$). Solutions of $f(z) = z$ are obtained by solving $z^2 - z + 1 = 0$, and these roots must also be solutions of $f(f(z)) = z = g(z)$ or of the quartic $z^4 + 2z^2 - z + 2 = 0$. It then follows that $z^2 - z + 1$ is a factor of $z^4 + 2z^2 - z + 2$. By direct division, the other factor is $z^2 + z + 2$ (details left to the reader), and the other two roots are the roots of $z^2 + z + 2 = 0$. Using the quadratic formula, the four roots (thus the desired fixed points) are $\frac{1}{2} \pm \frac{1}{2}\sqrt{3}i$ and $-\frac{1}{2} \pm \frac{1}{2}\sqrt{7}i$.

DYNAMIC SYSTEMS, FRACTALS, AND THE JULIA AND MANDELBROT SETS

A **dynamic system** is defined as a mapping from the complex plane into itself. Iterates of such mappings lead to the important applications in this area. G. Julia and P. Fatou were the early pioneers (1918), and by using topology, they developed detailed mathematical analysis for such systems. Some results involved what later became known as "fractals". You are no doubt familiar with Koch's curve, discovered in 1904—the classic example of a closed curve having infinite length and finite area, and possessing no tangents. It is the geometric counterpart to Weierstrass' function in analysis that is continuous but nowhere differentiable. However, without benefit of the computer, accurate drawings were not available at the time, and research in this area lay virtually dormant until recent times.

In 1975, Benoit Mandelbrot developed the theory further and later used modern computers to explore the geometric aspects. He coined the term *fractal*, and established programs to generate them on the computer, thus creating a popular area of computer art.

For our purpose, we define a **fractal** as any geometric object F in $\mathbb{R}^2$ that is *self-similar under dilation*. That is, any neighborhood of each point of F is a dilation of a part of F itself. One normally assumes also that such sets are to have *no tangents*. This means, for example, that the unit circle, although self-similar, is not a fractal.

In order to understand what is known as a *Julia set*, we must look at a few details involving iterates of dynamical systems. Let f be a polynomial in complex numbers, and define the **iterates** of f as follows:

$$f^1(z) = f(z), \quad f^2(z) = f[f(z)], \quad f^3(z) = f\{f[f(z)]\}, \quad \cdots$$

In general,

$$f^n(z) = f[f^{n-1}(z)]$$

where n is an integer ≥ 2. A point z is called **periodic** if for some $p \geq 1$, $f^p(z) = z$; its **period** is the least such integer p. The **orbit** of a point z is the set of complex numbers $\{f^1(z), f^2(z), f^3(z), \cdots, f^n(z), \cdots\}$. [The classic example is the function $f(z) = cz$, where c is a pth root of unity $\neq 1$, p prime. Here, $f^n(z) = c^n z$, and every point in the plane except the origin is periodic with period p; orbits of $c^n z$ (n a positive integer) lie on concentric circles centered at 0.] Finally, if w has period p and we define $g(z) = f^p(z)$, then g is also a dynamic system, and w is a *fixed point of $g(z)$*.

For fixed points of dynamical systems, it is important to consider what are called *attractive fixed* points and *repelling fixed* points. A fixed point w of g is called **attractive** under g iff $|g(z) - g(w)| < a|z - w|$ for all complex z in some neighborhood N of w and for some real constant $a < 1$. A fixed point w is called **repelling** if $|g(z) - g(w)| > a|z - w|$ for a real constant $a > 1$. Thus it follows that the sequence $z_n \equiv g^n(z) \rightarrow w$ as $n \rightarrow \infty$ for an attractive fixed point, and $z_n \rightarrow \infty$ if w is repelling (see Figure 9.13 for illustration). That is, the orbit under $g(z)$ of every point z within some neighborhood of w converges to w if w is attractive, and diverges to ∞ if w is repelling. Noting the two inequalities above, and the fact that the analyticity of g implies that $\left|[g(z) - g(w)]/[z - w]\right|$ converges to $|g'(w)|$ as $z \rightarrow w$, we have the following rule:

(9.6)   The point w is an attractive fixed point of g iff $|g'(w)| < 1$, and repelling iff $|g'(w)| > 1$. If $g'(w) = 1$, w can be either attractive or repelling.

Related to f, we call w an **attractive** (or **repelling**) **periodic point** of f iff w has period p under f for some positive integer p, and $|g'(w)| < 1$ (or $|g'(w)| > 1$) where $g(z) = f^p(z)$.

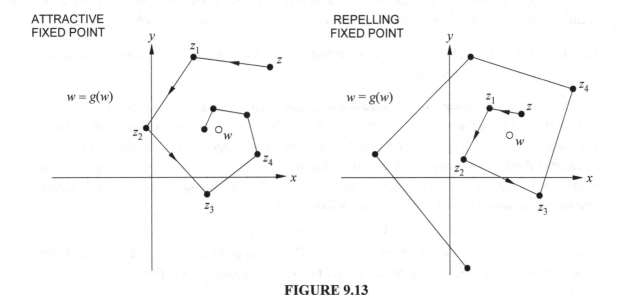

ATTRACTIVE
FIXED POINT

REPELLING
FIXED POINT

$w = g(w)$

$w = g(w)$

FIGURE 9.13

EXAMPLE 8

Consider the polynomial $g(z) = z^4 + 2cz^2 + c^2 + c \equiv (z^2 + c)^2 + c$, $c \neq 0$, which is the second iterate of $f(z) = z^2 + c$ The fixed points are the four solutions of $(z^2 + c)^2 + c = z$ which can be found using the quadratic formulas on the two equations

$$z^2 - z + c = 0 \qquad \text{and} \qquad z^2 + z + c + 1 = 0$$

(See Example 7 for the special case $c = 1$.) If $c = \frac{4}{25} = 0.16$, find the repelling and attractive fixed points for g.

SOLUTION

Setting $g(z) = z$ to find the fixed points, we obtain $g(z) = z^4 + 0.32z^2 + 0.1856 = z$ (exact decimals). So we must solve for the four solutions of the 4th degree equation $z^4 + 0.32z^2 - z + 0.1856 = 0$. But it is found that this equation has the factored form

$$(z^2 - z + 0.16)(z^2 + z + 1.16) = 0$$

(You are to work Problem 23 for the details.) From the quadratic equation we find the four fixed points

$$w_1 = \frac{4}{5}, \qquad w_2 = \frac{1}{5}, \qquad w_3 = \frac{-5 + \sqrt{91}i}{10}, \qquad w_4 = \frac{-5 - \sqrt{91}i}{10}$$

Observe that $g'(z) = 4z^3 + 0.64z = 4z(z^2 + \frac{4}{25})$. Thus we obtain $|g'(w_1)| = \frac{64}{25}$ and $|g'(w_2)| = \frac{4}{5}$. Since $|w_3| > 1$ and $|w_4| > 1$, we obtain $|g'(z)| = 4|z|(|z^2 + \frac{4}{25}|) \geq 4 > 1$ for $z = w_3, w_4$. Thus by **(9.6)** w_2 is an attractive fixed point, while w_1, w_3, and w_4 are repelling fixed points.

The Julia set corresponding to any analytic function can now be defined, using one further topological term in addition to those found in Chapter 5. Recall that a set S is said to be *closed* if and only if it contains all its boundary points. For arbitrary sets, we define the **closure** of S to be the set S itself plus all its boundary (or limit) points. For example, the closure of the open interval (a, b) on the x-axis is the closed interval $[a. b]$). Further examples: the closure of the open disk $|z - z_0| < r$ is the closed disk $|z - z_0| \leq r$. If S is the set of all rational points inside $|z - z_0| < r$ [that is, points having *rational coordinates*, like $(\frac{1}{2}, \frac{1}{4})$

and $(\frac{2}{3}, \frac{1}{3})$], the closure of S is again the closed disk $|z - z_0| \leq r$. It is a somewhat challenging exercise—seemingly trivial from the terminology—to prove that the *closure of any set is closed*.

DEFINITION   The **Julia set** $J(f)$ corresponding to an analytic function $f(z)$ is the closure of the set of all repelling periodic points of $f(z)$.

To make the discussion more manageable from this point on, we consider a special case. We assume that f is the quadratic function $f(z) = z^2 + c$, where c can be any (constant) complex number, denoted f_c (as in Example 8). The Julia set $J_c \equiv J(f_c)$ can be shown to be nonempty for each c (each iterate of f_c has a fixed point due to the fundamental theorem of algebra). The first iterate $f^1(z) = z^2 + c$ has two fixed points, the two roots of the quadratic equation $z^2 - z + c = 0$. The second iterate is a fourth degree polynomial, obtained by simple algebra (as in Example 8):

$$(z^2 + c)^2 + c = z^4 + 2cz^2 + c^2 + c$$

which has four fixed points. The third iterate is an 8th–degree polynomial, having eight fixed points, and so on. Theoretically, one can determine which of these fixed points are repelling, thus members of J_c.

The Julia sets J_c are among one of the most interesting collection of fractals. They can be displayed on the computer for various values of c. A few mathematical results independent of the computer are also interesting:

- J_c is a fractal for each complex $c \neq 0$.
- J_c is the unit circle $|z| = 1$ if $c = 0$ (which is not a fractal).
- For real c, $0 < c < \frac{1}{4}$, J_c is a closed curve having no tangents.
- J_c is symmetric with respect to the origin.
- If the orbit of c under f_c is bounded, J_c is connected. Otherwise, J_c is *totally disconnected* (i.e., the largest connected subset is a single point).

We now turn our attention to the famous Mandelbrot set corresponding to f_c (the function $f(z) = z^2 + c$). Interestingly enough, this set has intricate connections with the Julia sets J_c. Again we consider the iterates of f_c, this time *evaluated at $z = 0$*. For convenience, define the **Mandelbrot sequence** $\{c_n\}$ where $c_n = f_c^n(0)$ (with $c_1 = c$). Thus, the Mandelbrot sequence is just the orbit of $z = 0$ under f_c. By algebra, the sequence starts out $c, c^2 + c, (c^2 + c)^2 + c, \cdots$, and so on. In general this sequence is defined recursively by

$$c_1 = c, \quad c_{n+1} = c_n^2 + c \quad \text{for } n \geq 1$$

Various values of c lead to different Mandelbrot sequences. Here are some examples:

$c = 1$: $\{c_n\} = \{1, 2, 5, 26, 677, \cdots\}$

$c = -1$: $\{c_n\} = \{-1, 0, -1, 0, -1, \cdots\}$

$c = 1 + i$: $\{c_n\} = \{1 + i, 1 + 3i, -7 + 7i, -1 - 97i, -9408 - 193i, \cdots\}$

DEFINITION   The **Mandelbrot set** is the set M of all complex numbers c for which the corresponding Mandelbrot sequence $\{c_n\}$ is *bounded*. That is,

$$M = \{c : \text{there exists a real number } B \text{ such that } |c_n| \leq B \text{ for all integers } n\}$$

Thus, we see from the above examples that -1 belongs to M, but that 1 and $1 + i$ do not. Can you show that -2 belongs to M? Although it is difficult to obtain results using only algebra, a limited number of theorems can be proven fairly easily. Here are a few general results for the Mandelbrot sequence $\{c_n\}$:

(9.7) If $|c_n| \geq |c_1|$ for any particular value of n, then $|c_n|^2 - |c_n| \leq |c_{n+1}| \leq |c_n|^2 + |c_n|$.

(9.8) If the sequence $\{c_n\}$ converges, its limit is $\frac{1}{2}(1 \pm \sqrt{1-4c})$.

(9.9) M lies inside or on the circle $|z| = 2$.

Problem 25 establishes both **(9.7)** and **(9.8)**. One can use **(9.7)** to show that any member of M has magnitude ≤ 2: Suppose that $c \in M$ and $|c| > 2$. Then for some $\varepsilon > 0$, $|c| = 2 + \varepsilon$. By repeated use of **(9.7)** (with $c = c_1$) we obtain

$$|c_2| \geq |c_1|^2 - |c_1| = (2 + \varepsilon)^2 - (2 + \varepsilon) = 2 + 3\varepsilon + \varepsilon^2 > 2 + 3\varepsilon > |c_1|$$

$$|c_3| \geq |c_2|^2 - |c_2| \geq (2 + 3\varepsilon)^2 - (2 + 3\varepsilon) = 2 + 9\varepsilon + 9\varepsilon^2 > 2 + 9\varepsilon > |c_1|$$

and in general (by mathematical induction) $|c_n| > 2 + 3^n \varepsilon$. Hence, the sequence $\{c_n\}$ is unbounded, contradicting that c was a member of M. Thus, $|c| \leq 2$.

The Mandelbrot set is a fractal having many unusual features. It is illustrated in Figure 9.14. We

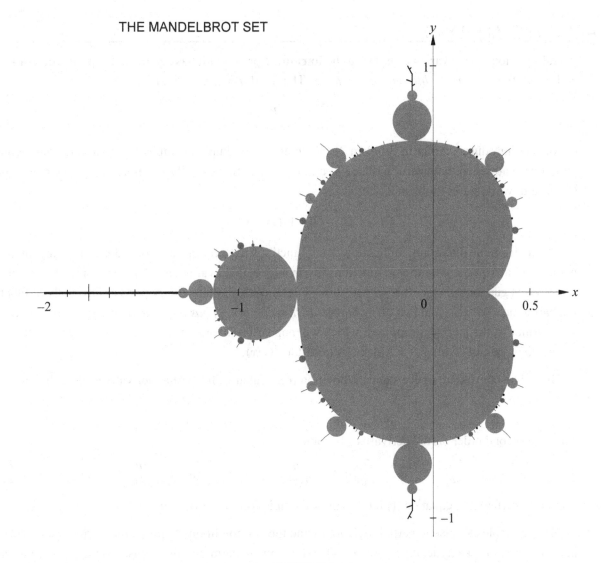

THE MANDELBROT SET

FIGURE 9.14

mention here just a few interesting results concerning the Mandelbrot set; some of these require advanced mathematical analysis (see, for example, Barnsley and Bélair listed in the bibliography).

- *M* is symmetric about the *x*-axis.
- *M* is connected.
- *M* meets the real axis on the interval $[-2, \frac{1}{4}]$.
- If *c* is a member of *M* in its main cardioid, then J_c is homeomorphic to a circle (it is the image the unit circle under a one-to-one continuous function with continuous inverse).

The Julia sets J_c for arbitrary *c* are intimately tied to the Mandelbrot set by the following result (due to H. Brolin):

THEOREM 9.10

A point *c* is a member of the Mandelbrot set *M* iff the Julia set J_c is connected.

HARMONIC ANALYSIS

A real function of two variables is said to be **harmonic** provided it has second order partial derivatives and it satisfies *LaPlace's differential equation*. That is, if $u = u(x, y)$, then

$$\frac{\partial^2 u}{\partial x^2} + \frac{\partial^2 u}{\partial y^2} = 0$$

(a concept introduced earlier in Problem 28, Chapter 3). A simple example is $u = xy$; since the second partial derivatives are both zero, LaPlace's equation is satisfied trivially. A more complicated example (for you to work out in Problem 27 below) is

$$u = \ln(x^2 + y^2)$$

Aside from the important implications such functions have for applications, there are some interesting mathematical aspects of harmonic functions involving complex variables. An abundance of harmonic functions is guaranteed since, as we will show, both the real and imaginary parts of any analytic function are harmonic. We also find that *every harmonic function u(x, y) has a harmonic conjugate v(x, y) that is also harmonic*, the precise meaning of which will appear in Theorem 9.11 below. For example, the harmonic conjugate of $\frac{1}{2}\ln(x^2 + y^2)$ happens to be $\tan^{-1}(y/x)$.

It will be convenient to use standard notation from calculus for partial derivatives. Recall that

$$u_x = \frac{\partial u}{\partial x} \qquad \text{and} \qquad u_y = \frac{\partial u}{\partial y}$$

Thus the second order partial derivatives become

$$u_{xx} = \frac{\partial}{\partial x}(u_x) = \frac{\partial^2 u}{\partial^2 x}, \qquad u_{yx} = \frac{\partial}{\partial y}(u_x) = \frac{\partial^2 u}{\partial y \partial x}, \qquad u_{xy} = \frac{\partial}{\partial x}(u_y) = \frac{\partial^2 u}{\partial x \partial y}, \qquad \text{and} \qquad u_{yy} = \frac{\partial^2 u}{\partial y^2}$$

LaPlace's differential equation then takes on the simple form $u_{xx} + u_{yy} = 0$.

It is a simple exercise in partial derivatives and the CR conditions to prove that if $f(z) = f(x + iy) = u(x, y) + iv(x, y)$ is analytic, then both $u(x, y)$ and $v(x, y)$ are harmonic functions. The CR conditions are

$$u_x = v_y \qquad \text{and} \qquad u_y = -v_x$$

Thus,

$$u_{xx} = v_{xy} \quad \text{and} \quad u_{yy} = -v_{yx}$$

Because f is analytic, $f''(z)$ exists and is continuous, which implies that v_{xy} and v_{yx} are continuous [via (3.7)]. It then follows that $v_{xy} = v_{yx}$ and

$$u_{xx} + u_{yy} = v_{xy} - v_{yx} = 0$$

An almost identical argument shows that $v(x, y)$ also satisfies LaPlace's equation, proving that both u and v are harmonic. The proof of the first half of the following theorem has thus been established; the proof of the converse involves advanced analysis and will not be given here.

THEOREM 9.11

If $f(x + iy) = u(x, y) + iv(x, y)$ is analytic, then both $u(x, y)$ and $v(x, y)$ are harmonic functions. Conversely, if $u(x, y)$ has continuous second order partial derivatives on a simply-connected region R and is harmonic, then a second harmonic function $v(x, y)$ exists, called the **harmonic conjugate** of $u(x, y)$, such that $u(x, y) + iv(x, y)$ is analytic on R.

It is often possible to construct the harmonic conjugate from a given harmonic function by an elementary procedure, as illustrated by the following example.

EXAMPLE 9

Show that $u = 3x^2y - y^3$ is harmonic, and find its harmonic conjugate.

SOLUTION
Computing the partial derivatives, we have

$$u_x = 6xy \quad \text{and} \quad u_{xx} = 6y$$
$$u_y = 3x^2 - 3y^2 \quad \text{and} \quad u_{yy} = -6y$$

Hence,

$$u_{xx} + u_{yy} = 6y - 6y = 0$$

To find the harmonic conjugate, we want a function $v(x, y)$ that satisfies the CR conditions:

$$v_y = u_x, \quad v_x = -u_y$$

Thus $v_y = 6xy$ and

$$v = \int 6xy\,dy = 3xy^2 + h(x)$$

where $h(x)$ is a function of x alone—the constant of integration with respect to y. To determine $h(x)$, take the partial derivative of v with respect to x:

$$v_x = 3y^2 + h'(x) \quad \Rightarrow \quad 3y^2 + h'(x) = -u_y = -3x^2 + 3y^2 \quad \Rightarrow \quad h'(x) = -3x^2$$

Hence $h(x) = -x^3 + C$ (C constant). To make things simple, let $C = 0$. Thus we find the harmonic conjugate of u:

$$v = 3xy^2 - x^3$$

[You might notice that if $f(z) = u + iv = 3x^2y - y^3 + i(-x^3 + 3xy^2)$, then $f(z) = -iz^3$, which is, of course, analytic.]

ANALYTIC CONTINUATION AND MULTIPLE-VALUED FUNCTIONS

Another interesting concept in complex analysis is known as **analytic continuation**. Recall that, by definition, a function is analytic at a point iff it is differentiable in a neighborhood of that point, hence has a Taylor series representation at that point. Thus if $f(z)$ is analytic in an open region R, and is analytic at some boundary point z_0 of R, then the neighborhood N of analyticity centered at z_0 extends beyond R (see Figure 9.15). It follows that $f(z)$ is defined in a larger region having points both in R and points not in R. In this case, f (as defined on N) is said to be an *analytic continuation* of f in R. Thus, we are provided with additional information about f outside R.

More generally, suppose that a function $f_1(z)$ is analytic on an open connected region R_1, and that another function $f_2(z)$ is analytic on an open connected region R_2 different from R_1 that meets R_1 in at least one point (that is, $R_1 \cap R_2$ is nonempty), as illustrated in Figure 9.16. Then if $f_2(z) = f_1(z)$ on $R_1 \cap R_2$, f_2 is called an **analytic continuation** of f_1. We can immediately see the symmetry of this relation: if f_2 is an analytic continuation of f_1, then f_1 is an analytic continuation of f_2.

ANALYTIC CONTINUATION

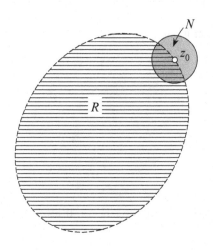

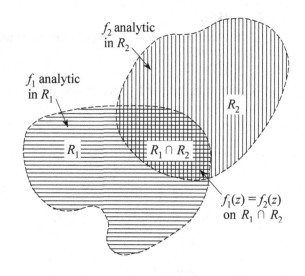

FIGURE 9.15   **FIGURE 9.16**

EXAMPLE 10

Consider the two functions defined by power series

$$f(z) = \sum_{n=0}^{\infty} z^n = 1 + z + z^2 + z^3 + \cdots, \quad |z| < 1$$

and

$$g(z) = i \sum_{n=0}^{\infty} \left(\frac{z-c}{-i} \right)^n = i - (z-c) - i(z-c)^2 + \cdots, \quad |z-c| < 1$$

where $c = 1 + i$. Show that g is an analytic continuation of f.

SOLUTION

The first series is a geometric series, with $f(z) = (1 - z)^{-1}$ for $|z| < 1$. The second series is also a geometric series, valid for $|z - c| < 1$. Since $|(z - c)/(-i)| = |z - c| < 1$ we obtain

$$g(z) = i \sum_{n=0}^{\infty} \left(\frac{z - c}{-i}\right)^n = \frac{i}{1 - \frac{z - c}{-i}} = \frac{i(-i)}{-i - (z - c)} = \frac{1}{-i - (z - 1 - i)} = \frac{1}{1 - z}$$

If we let R_1 be the interior of the circle $|z| = 1$ and R_2 that of $|z - c| = 1$, as illustrated in Figure 9.17, then $f(z) = g(z)$ on the intersection $R_1 \cap R_2$. Thus by definition, $g(z)$ is an analytic continuation of $f(z)$.

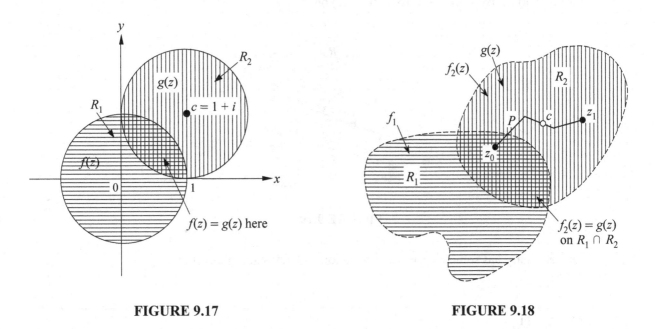

FIGURE 9.17   **FIGURE 9.18**

A very important theorem is the following; it is a key result for a few later ideas.

THEOREM 9.12

The analytic continuation of a function is unique.

> *Proof:* (1) Suppose f_1 on R_1 has an analytic continuation f_2 on R_2, and that g is another analytic continuation of f_1 also defined on R_2 (Figure 9.18). Thus $g(z) = f_1(z) = f_2(z)$ for all z in a neighborhood of a point z_0 in $R_1 \cap R_2$. We must prove that if z_1 is any point in R_2, $g(z_1) = f_2(z_1)$. Let P be a polygonal path in R_2 joining z_0 and z_1. Since g and f_2 are analytic at z_0 and $g(z_0) = f_2(z_0)$, the derivatives of all orders of g and f_2 agree at z_0. By Taylor's theorem, the Taylor series representations of g and f_2 in powers of $z - z_0$ are identical in a neighborhood of z_0. Thus on some closed interval $[z_0, c]$ of P ($c > z_0$), $g(z) = f_2(z)$.
>
> (2) Now we move c as far as we can on P towards z_1, while maintaining equality between g and f_2. Thus we obtain a maximal interval $[z_0, c]$ for which equality holds. (The right endpoint of this interval is the least upper bound of the set of all c on P for which equality holds on $[z_0, c]$.) Suppose that $c < z_1$ on P. As before, since g and f_2 are analytic at c, the Taylor series representations of g and f_2 at c are identical on a neighborhood of c, thus for all z in $[z_0, d]$ for some $d > c$. Hence $g(z) = f_2(z)$ on this interval. But this contradicts our definition of $[z_0, c]$ as the largest such interval. Therefore, $c = z_1$ and $g(z_1) = f_2(z_1)$, as desired. ⬙

COROLLARY

Suppose that $f(z)$ and $g(z)$ are analytic functions on an open connected region R. If f and g agree on an arc C of a curve inside R, then $f(z) = g(z)$ on R.

Proof: Let c be any point of C, and let $h(z) =$ Taylor series expansion of $f(z)$ about $z = c$, valid in some neighborhood N of c (see Figure 9.19). The derivatives of f and g can be obtained by taking limits of difference quotients as z approaches c on C. Since $f(z) = g(z)$ on C, the derivatives of g at c are identical with those of f. Hence the Taylor series expansion of $g(z)$ in N coincides with that of $f(z)$, so that $g(z) = h(z)$ on N. Hence, f and g are both analytic continuations of h from N to R. By Theorem 9.12, $f(z) = g(z)$ on R. ＼

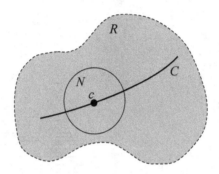

FIGURE 9.19

This corollary has some surprising implications. A few examples follow.

EXAMPLE 11

An analytic function $f(z)$ in a connected open region R can have a single zero z_0 (or even infinitely many of them) in a region R without being identically zero, but if z_0 is the endpoint of an arc in R, however short, and $f(z) = 0$ on that arc, then by the corollary above, $f(z) = 0$ on the entire region R.

EXAMPLE 12

Since any closed interval on the real axis is an arc in any region containing that interval, an identity valid for real numbers on the real axis extends to its counterpart in complex variables. For example, since $\sin^2 x + \cos^2 x = 1$ for all real x, then $\sin^2 z + \cos^2 z - 1 = 0$ on the x-axis (which contains an arc) and by the result of Example 11, $\sin^2 z + \cos^2 z - 1 = 0$ for all complex numbers z.

EXAMPLE 13

If $f(z)$ is an analytic function that agrees with $\sin x$ for real $z = x$, then $f(z)$ must have the form $\sin z = \sin x \cosh y + i \cos x \sinh y$, where $z = x + iy$. Similarly, if $f(z)$ is an analytic function that agrees with $\ln x$ for real $z = x > 0$, then $f(z) = f(re^{i\theta}) = \ln r + i\theta$.

The existence of singularities can obstruct analytic continuation. In Example 10 we observed the continuation of $f(z)$ to $g(z)$ where a singularity occurred at the point $z = 1$ for both functions. As a matter of fact, the single function represented by $(1 - z)^{-1}$ is the analytic continuation of both series $f(z)$ and $g(z)$ from R_1 and R_2 to the entire deleted plane R, $z \neq 1$. But the singularity $z = 1$ cannot be avoided: no analytic continuation of either $f(z)$ or $g(z)$ to an open set containing $z = 1$ exists. But $f(z)$ can be continued beyond R_1, as we saw in Example 10, and the boundary of R_1 is a circle containing 1. On the other hand, consider the series

$$h(z) = \sum_{n=1}^{\infty} nz^n , \qquad |z| < 1$$

This defines a function that cannot be analytically continued beyond the circle $|z| = 1$. This function (like that of $f(z)$ in Example 10) has points of singularity on the unit circle, but, unlike $f(z)$, $h(z)$ diverges for all points on $|z| = 1$ (since $n(\cos n\theta + i\sin n\theta) \nrightarrow 0$ as $n \rightarrow \infty$). Another way to look at it is that $f(z)$ has *removable singularities* for all z on the unit circle, except at $z = 1$, while $h(z)$ has none. The unit circle is called a **natural barrier** for analytic continuation of $h(z)$.

THEOREM 9.13

Suppose $f(z)$ is analytic on an open connected region R, having non-removable singularities on the boundary of R. If these singularities of $f(z)$ are dense (that is, every neighborhood of every point of the boundary contains a singular point), then there is no analytic continuation of $f(z)$.

EXAMPLE 14

The function $f(z) = \ln z$ as defined in Chapter 2 is defined but not continuous for $z = x \leq 0$ (therefore not analytic), but it is analytic at all other points in the complex plane. Let R_1 be this region, as illustrated in Figure 9.20). This is an open connected set since every point is an interior point of the set. But $\ln z$ cannot be analytically continued to include points on the negative x-axis. Prove this.

SOLUTION

If an open region R_2 contains points on the negative real axis then it contains points above and below it. Any analytic function $g(z)$ on this region, in order to be an analytic continuation of $\ln z$, must agree with $\ln z$ on $R_1 \cap R_2$. But this would make $g(z)$ discontinuous on the negative x-axis, a contradiction.

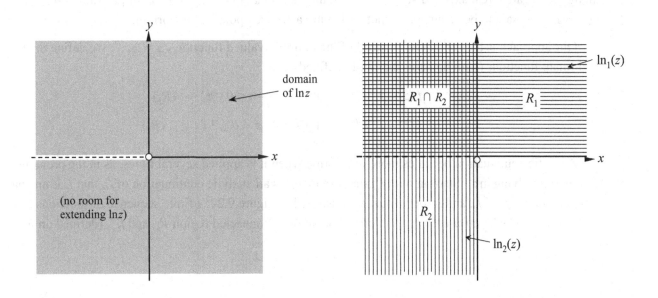

FIGURE 9.20   **FIGURE 9.21**

On the other hand, in spite of the problem just mentioned, a logarithm-like function (whose inverse equals e^z) does have an analytic continuation that includes $z = x \le 0$. One need only restrict the domain of $\ln z$ to make room for such an extension, which, in effect, defines a new function. Suppose we agree that instead of the domain of $\ln z$ extending all the way from $\text{Arg} z = -\pi$ counterclockwise to $\text{Arg} z = \pi$, we delete the lower half-plane and define a new domain R_1. That is, with $z = re^{i\theta}$, the new function is

$$\ln_1 z = \ln r + i\theta, \qquad r > 0, \, 0 < \theta < \pi \qquad (R_1)$$

Now consider a function with the same formula for computation, but having a different domain, overlapping R_1, as shown in Figure 9.21:

$$\ln_2 z = \ln r + i\theta, \qquad r > 0, \, \pi/2 < \theta < 3\pi/2 \qquad (R_2)$$

Both functions satisfy the equation $e^w = z$, where $w = \ln_k z$ ($k = 1, 2$), and are analytic. Since $\ln_1 z = \ln_2 z$ over the common region $\pi/2 < \theta < \pi$, $\ln_2 z$ *is the analytic continuation of* $\ln_1 z$. By continuing this process forward and backward ad infinitum, one obtains the traditional logarithm function for complex variables, $\ln z = \ln r + i\theta$, where θ is arbitrary ($z = re^{i\theta}$). Thus $\ln z$ has an infinity of values by taking different values for θ.

Such functions are traditionally called *multiple-valued functions*, a misnomer since in modern mathematical theory, there is no such thing. One might say that the inverse of the real function $\sin x$ is "multiple valued", but the resulting confusion is avoided by defining the legitimate function $\sin^{-1} x$, as in calculus. The traditional term arises from the ambiguity created by confusing the *solutions* of an equation (used to define a "function") with the *values of the imagined inverse* "function". This is not to say that the early pioneers in complex variables were incorrect because of their terminology. They fully realized a fundamental problem here and they dealt with it by various means. The concept of *Riemann surfaces* was invented to clarify the meaning, an area that has been thoroughly researched. And there is something to be said in favor of a less complicated concept for analyticity when this "incorrect" definition is allowed.

Another example besides $\ln z$ is the function $\sqrt{z}$, defined for all complex z (as in Chapter 2). The solutions of the equation $w^2 = z$ on the other hand consist of the *two* values $\pm\sqrt{z}$ for each z. The function $f(z) = \sqrt{z}$ defined in Chapter 2 is a legitimate analytic function mapping $\mathbb{C}$ into $\mathbb{C}$ having singularities along the negative real axis and at $z = 0$. But in the traditional treatment, the "multiple-valued function" $g(z) = \pm\sqrt{z}$ is said to be an analytic function having a "branch point" at the origin.

Our approach is the following. Instead of the multiple-valued function $\sqrt{z} = \pm z^{1/2}$, we define the *two* legitimate functions (with domains explicitly defined):

$$f_1(z) = r^{1/2} e^{i\theta/2} \qquad \text{for } r > 0, \, -\pi < \theta \le \pi^{\dagger} \qquad (R_1)$$

$$f_2(z) = r^{1/2} e^{i\theta/2} \qquad \text{for } r > 0, \, \pi < \theta \le 3\pi \qquad (R_2)$$

Analytic continuation becomes more interesting when it is applied several times, as in the previous example involving $\ln z$. Starting with f_1, suppose that f_2 is an analytic continuation of f_1, that f_3 is an analytic continuation of f_2, and so on. Thus, as illustrated in Figure 9.22, a finite sequence of functions $f_1, f_2, f_3, \cdots, f_n$ results, where for each k, f_k is defined on an open connected region R_k, and f_{k+1} (defined on R_{k+1})

---

†This is the square root radical defined in Chapter 2.

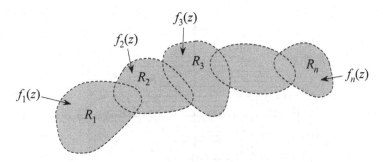

FIGURE 9.22

is an analytic continuation of f_k ($k = 1, 2, \cdots$). One then defines a function $F(z)$ such that $F(z)$ has the value $f_k(z)$ on R_k. The domain of F is the entire collection of regions R_k (their union), and f_k is called an **element** of F.

The logarithm and square root functions can be used to illustrate an interesting phenomenon. First, suppose the function $f_1(z)$ has been analytically continued from its domain R_1 via the functions $f_2(z), f_3(z), f_4(z), \cdots, f_n(z)$ on the respective domains $R_2, R_3, R_4, \cdots, R_n$ (Figure 9.22). Suppose also that $f_1(z)$ has been analytically continued via the functions $g_2(z), g_3(z), g_4(z), \cdots, g_m(z)$ on $R'_2, R'_3, \cdots R'_m$ along a different path path, as shown in Figure 9.23. If R_n meets R'_m, $g_m(z)$ *need not agree with* $f_n(z)$ on their common domain $R_n \cap R'_m$. When this happens, it can be shown that there must exist a singular point of the function generated by the elements $f_k(z)$ and $g_k(z)$ lying between the two paths.

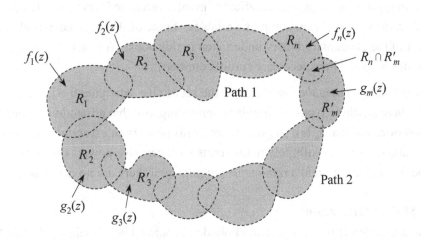

FIGURE 9.23

The functions $\ln_k z$ for $k = 1, 2$ defined in Example 14, where $\ln_2 z$ is the analytic continuation of $\ln_1 z$ from R_1 to R_2 can be used to provide an illustration. Suppose one further defines $\ln_3 z$ and $\ln_4 z$ as $\ln r + i\theta$, $r > 0$, for $\pi < \theta < 2\pi$ (R_3) and for $3\pi/2 < \theta < 5\pi/2$ (R_4). (See Figure 9.24.) Then the two chains of analytic continuations $\{f_2(z), f_1(z)\}$ and $\{f_2(z), f_3(z), f_4(z)\}$ lead to two functions $\ln_1(z)$ and $\ln_4(z)$ having a domain in common (the set $R_1 \cap R_4$). But $\ln_1 \neq \ln_4$ on $R_1 \cap R_4$ (observe: $\ln_1(1 + i) = \ln\sqrt{2} + \pi i/4$, $\ln_4(1 + i) = \ln\sqrt{2} + 9\pi i/4$).

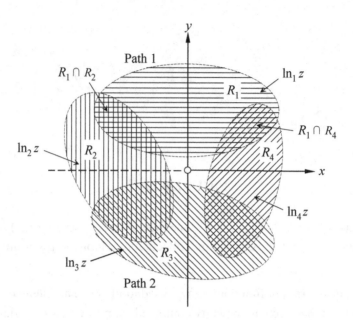

FIGURE 9.24

A FEW INTERESTING THEOREMS

Many interesting theorems in complex variables involve advanced concepts. They show the rich variety of mathematical ideas possible in complex variables. Some of them are nevertheless easy to understand in the context of the elementary theory undertaken in this book. We include a discussion of a few of these results, with definitions where necessary (proofs omitted).

Recall that a *deleted neighborhood* of z_0 is the set of all z satisfying $0 < |z - z_0| < r$ for some positive real r. Also, there are three types of singular points. Suppose that a function is analytic in some deleted neighborhood of z_0 and that z_0 is a singular point. This point is an *essential singularity* if the Laurent series of $f(z)$ about $z - z_0$ has infinitely many terms with negative powers of $z - z_0$, a *pole* if the series has a finite number of such terms, and a *removable singularity* if there are no negative powers of $z - z_0$.

THEOREM 9.14 (Riemann)

Suppose that a function $f(z)$ is analytic and bounded in some deleted neighborhood of z_0. Then $f(z)$ is either analytic at z_0, or z_0 is a removable singularity.

THEOREM 9.15 (Casorati-Weierstrass)

If $f(z)$ is analytic in some deleted neighborhood N of z_0, but has an essential singularity at z_0, then as z varies on N, $f(z)$ comes arbitrarily close to any given complex number.

EXAMPLE 15

Consider the analytic function $e^{1/z}$ which has an essential singularity at $z = 0$. Consider the extremely small neighborhood $N : |z| < 10^{-20}$. Observe that if $z \in N$ and $z = x$ (real), then

$$f(z) = e^{1/x} \approx e^{10^{20}}$$

Thus, $|f(z)|$ becomes extremely large as $\Re z \to 0$ on N. It would seem from this that it would be impossible to choose z in this neighborhood so that $f(z) = \frac{1}{2}$ or 0, for example. But by the Casorati-Weierstrass theorem, there exists a sequence $z_n \to 0$ in N for which $f(z_n) \to 0.01$, one for which $f(z_n) \to -0.001$ and one for which $f(z_n) \to 10^{-100} + 100i$. Moreover, there is also a sequence $z_n \to 0$ for which $f(z_n) \to 10^{1000}(2 - 3i)$.

THEOREM 9.16 (Rouche)

Let two functions $f(z)$ and $g(z)$ be analytic on the interior of a simple closed curve C and on C itself, and suppose that $|f(z)| > |g(z)|$ on C. Then, counting multiplicities, $f(z)$ and $f(z) + g(z)$ have the same number of zeroes inside C.

EXAMPLE 16

Consider the polynomial equation

$$z^9 - 2z^6 + 15z^3 + z + 1 = 0$$

Use Rouche's Theorem to show that this equation has three roots (counting multiplicities) inside the circle $|z| = \frac{1}{2}$.

SOLUTION

Let $f(z) = 15z^3$ and $g(z) = z^9 - 2z^6 + z + 1$. On the circle $|z| = \frac{1}{2}$, we obtain $|f(z)| = 15|z|^3 = \frac{15}{8}$, while $|g(z)| = |z^9 - 2z^6 + z + 1| \le |z|^9 + 2|z|^6 + |z| + 1 = (\frac{1}{2})^9 + 2(\frac{1}{2})^6 + \frac{1}{2} + 1 \approx 1.533 < \frac{15}{8}$, and the inequality $|f(z)| > |g(z)|$ holds on $|z| = \frac{1}{2}$. Hence, by Rouche's theorem, the polynomial equation $f(z) + g(z) = 0$ has the same number of roots as $f(z) = 0$ inside $|z| = \frac{1}{2}$. Since $z = 0$ is a root of $15z^3 = 0$ having multiplicity 3, the given equation above has 3 roots inside $|z| = \frac{1}{2}$, counting multiplicities.

The next theorem was discovered independently by two mathematicians, H.A. Schwarz (1843–1921) and E.B. Christoffel (1829–1900).

THEOREM 9.17 (Schwarz-Christoffel Transformation)

Given a polygon P in the complex plane, there exists a transformation $w = f(z)$ that maps the real axis one-to-one onto P, where f is defined and continuous in the upper-half plane $y \ge 0$, and conformal except at the points on the x-axis that map to the vertices of P.

EXAMPLE 17

It can be shown that the mapping $w = f(z)$, where

$$f(z) = \int_1^z (w + 1)^{-2/3}(w - 1)^{-2/3}\, dw$$

maps the real axis (including the point at infinity) onto the equilateral triangle having vertices 0, 1, and $\frac{1}{2} + \frac{\sqrt{3}}{2} i$.

Closely related to this is *Riemann's mapping theorem*, which essentially states that given any simple closed curve C in the complex plane, there exists an analytic function that maps the unit circle onto C. The next theorem was discovered by G. Mittag-Leffler (1846–1927), a student of Weierstrass.

THEOREM 9.18 (Mittag-Leffler)

Suppose that $f(z)$ is an analytic function throughout the complex plane, except for simple poles $a_1, a_2, a_3,$ $\cdots$ (arranged in order of non-decreasing absolute values), and that $b_1, b_2, b_3, \cdots$ are the residues of $f(z)$ at $a_1, a_2, a_3, \cdots$. Furthermore, suppose there exist circles C_N centered at the origin having radius R_N not passing through any poles, such that $|f(z)|$ has a fixed bound M on all these circles, and that $R_N \to \infty$ as $N \to \infty$. Then

(9.18)
$$f(z) = f(0) + \sum_{n=0}^{\infty} b_n \left(\frac{1}{z - a_n} + \frac{1}{a_n} \right)$$

EXAMPLE 18

Apply the Mittag-Leffler theorem to the function $f(z) = \cot z - \dfrac{1}{z}$ and obtain a series expansion of the form (9.18).

SOLUTION

Since $f(z) = \dfrac{z \cos z - \sin z}{z \sin z}$, this function is analytic at all points except at the zeroes of $z \sin z$. The nonzero solutions of $z \sin z = 0$ are simple poles, occurring at $z = n\pi$, where $n = \pm 1, \pm 2, \pm 3, \cdots$ ($z = 0$ is a removable singularity). So we let $a_1 = \pi$, $a_2 = -\pi$, $a_3 = 2\pi$, $a_4 = -2\pi$, $a_5 = 3\pi$, and so on. To calculate the residues, we use (7.11):

$$b_n = \lim_{z \to n\pi} (z - n\pi) \left(\frac{z \cos z - \sin z}{z \sin z} \right) = \lim_{z \to n\pi} \left(\frac{z - n\pi}{\sin z} \right) \cdot \lim_{z \to n\pi} \left(\frac{z \cos z - \sin z}{z} \right)$$

By L'Hospital's rule,

$$b_n = \lim_{z \to n\pi} \left(\frac{1}{\cos z} \right) \cdot \lim_{z \to n\pi} \left(\frac{z \cos z - \sin z}{z} \right) = \frac{1}{\cos n\pi} \cdot \frac{n\pi \cos n\pi - \sin n\pi}{n\pi} = 1$$

Because $f(z)$ converges to zero as $z \to 0$, $z = 0$ is a removable singularity and we can take $f(0) = 0$. Now consider the circles $C_N : |z| = (2N + 1)\pi/2$ for $N = 1, 2, 3, \cdots$. We find that for any N, if z lies on C_N, then

$$|f(z)| = \left| \cot z - \frac{1}{z} \right| \leq |\cot z| + \frac{1}{|z|} < |\cot z| + 1$$

To show that $f(z)$ has a fixed bound on all the C_n's, recall that the analysis in Example 11, Chapter 8, is precisely what we need here. We established there that $|\cot z| < B$ for $z \in C_k$ [radius $(2k + 1)\pi/2 = R_N$]. Hence $|\cot z - 1/z| < B + 1$ for all $z \in C_N$. Thus, all hypotheses for (9.18) are satisfied, and

$$\cot z + \frac{1}{z} = \sum_{n=1}^{\infty} \left(\frac{1}{z - a_n} + \frac{1}{a_n} \right) \qquad \text{where } a_1 = \pi,\ a_2 = -\pi,\ a_3 = 2\pi,\ a_4 = -2\pi,\ a_5 = 2\pi,\ \cdots$$

To go a little further, suppose the individual terms of the series are spelled out. The series becomes

$$\left(\frac{1}{z - \pi} + \frac{1}{\pi} \right) + \left(\frac{1}{z + \pi} + \frac{1}{-\pi} \right) + \left(\frac{1}{z - 2\pi} + \frac{1}{2\pi} \right)$$

$$+ \left(\frac{1}{z + 2\pi} + \frac{1}{-2\pi} \right) + \left(\frac{1}{z - 3\pi} + \frac{1}{3\pi} \right) + \left(\frac{1}{z + 3\pi} + \frac{1}{-3\pi} \right) + \cdots$$

and therefore

(9.19)
$$\cot z - \frac{1}{z} = \sum_{n=1}^{\infty} \left(\frac{1}{z - n\pi} + \frac{1}{z + n\pi} \right) = \sum_{n=1}^{\infty} \frac{2z}{z^2 - (n\pi)^2}$$

Another interesting area is that of *infinite products* (analogous to infinite series). This area was developed by Euler and has many intriguing features. An infinite product has the form

$$f_1(z)f_2(z)f_3(z)\cdots \equiv \prod_{n=1}^{\infty} f_n(z)$$

Such a product is said to **converge** [**diverge**] if and only if the sequence of **partial products**

$$p_n = \lim_{k\to\infty} f_1(z)f_2(z)f_3(z)\cdots f_n(z) \equiv \lim_{n\to\infty} \prod_{k=1}^{n} f_k(z)$$

converges [**diverges**]. The **value** of a convergent infinite product is the limit of its partial products.

NOTE: This definition does not always produce what one's intuition might predict. For example, even though it might seem that $\frac{1}{2}\cdot\frac{2}{3}\cdot\frac{3}{4}\cdot\frac{4}{5}\cdots = 1$ (by simple cancellation), the definition above produces the answer zero (the partial product $p_n = \frac{1}{2}\cdot\frac{2}{3}\cdot\frac{3}{4}\cdot\frac{4}{5}\cdots\frac{(n-1)}{n} = \frac{1}{n} \to 0$ as $n\to\infty$). Because of this and related theoretical issues, from now on we adopt the traditional viewpoint that an infinite product *diverges* if its limit does not exist, or *if its limit is zero*. This convention allows us to prove a necessary condition for convergence for infinite products that is analogous to *infinite series*: If the *infinite product* $\prod a_n$ converges, then $\lim a_n = 1$ as $n\to\infty$ (consider the ratio p_n/p_{n-1} as $n\to\infty$). ⬊

The earliest treatments of infinite products consisted of applications of elementary algebra. One example is an identity due to Euler:

$$(1 + z)(1 + z^2)(1 + z^3)(1 + z^4)\cdots = \frac{1}{(1-z)(1-z^3)(1-z^5)(1-z^7)\cdots}$$

This formula may seem a bit odd in terms of its form, thus not immediately transparent. But Euler's argument is quite clear, going something like this:

Let P be the given product involving the factors $(1 + z^n)$ for positive integers n, and let Q be the corresponding product of factors of the form $(1 - z^n)$. Then, by elementary algebra (using the identity $(1 + w)(1 - w) = 1 - w^2$),

$$P = \frac{PQ}{Q} = \frac{(1-z^2)(1-z^4)(1-z^6)(1-z^8)(1-z^{10})\cdots}{(1-z)(1-z^2)(1-z^3)(1-z^4)(1-z^5)(1-z^6)(1-z^7)\cdots}$$

$$= \frac{1}{(1-z)(1-z^3)(1-z^5)(1-z^7)\cdots} \quad ⬊$$

How much Euler was aware of the pitfalls of such arguments is not known, but his use of elementary methods and intuitive reasoning to obtain important results free of error was a mark of genius. His outstanding work in infinites series was to a large extent done in this manner. His ability to make mental calculations involving large numbers was phenomenal. It was Euler who introduced the symbol i for $\sqrt{-1}$, and the formula **(2.5)** in Chapter 2 is due to Euler. By using intuitive reasoning, Euler deduced the exact value $\pi^2/6$ for the squared harmonic series, which had previously eluded many able mathematicians (we obtained this result earlier by using residues). Euler's work filled no less than 75 volumes, and it took 47 years after his death for the St. Petersburg Academy to finish printing all his manuscripts (Burton, 1991).

An elementary introduction to this area can be had by starting with a product of complex numbers like

$$(1 + a)(1 + b)(1 + c) = 1 + a + b + c + ab + ac + bc + abc$$

and generalizing this to the infinite product

$$\prod_{n=1}^{\infty}(1+a_n) = 1 + \sum_{n=1}^{\infty} a_n + \sum_{n \neq m}^{\infty} a_n a_m + \sum_{n \neq m \neq k}^{\infty} a_n a_m a_k + \cdots$$

A more useful approach is to use a basic identity from logarithms to convert a product into a sum:

$$\ln[(1+a_1)(1+a_2)(1+a_3)\cdots(1+a_n) = \ln(1+a_1) + \ln(1+a_2) + \ln(1+a_3) + \cdots + \ln(1+a_n)$$

which proves

$$\ln \prod_{n=1}^{\infty}(1+a_n) = \sum_{n=1}^{\infty} \ln(1+a_n) \qquad \text{and} \qquad \prod_{n=1}^{\infty}(1+a_n) = \exp \sum_{n=1}^{\infty} \ln(1+a_n)$$

Thus we can see that the infinite product $\prod(1+a_n)$ converges iff the series $\sum \ln(1+a_n)$ converges. [Note that one can always convert a product of the form $\prod b_n$ to the form $\prod(1+a_n)$ by the substitution $a_n = b_n - 1$. It is often more convenient to use the form involving $1 + a_n$.]

It is considerably easier to prove results for infinite products of real numbers than those for complex products. A few numerical results (listed below) illustrate the range of ideas possible for such products. All these results will eventually be established.

$$\frac{3}{4} \cdot \frac{8}{9} \cdot \frac{15}{16} \cdots \frac{n^2-1}{n^2} \cdots = \frac{1}{2} \qquad (n \geq 2)$$

$$\frac{4}{3} \cdot \frac{36}{35} \cdot \frac{100}{99} \cdots \frac{(4n-2)^2}{(4n-2)^2-1} \cdots = \sqrt{2}$$

$$\frac{8}{9} \cdot \frac{24}{25} \cdot \frac{48}{49} \cdots \frac{(2n+1)^2-1}{(2n+1)^2} \cdots = \frac{\pi}{4} \qquad \text{(a form of Wallis' product)}$$

$$\frac{8}{9} \cdot \frac{80}{81} \cdot \frac{224}{225} \cdots \frac{(6n-3)^2-1}{(6n-3)^2} \cdots = \sqrt{3}/2$$

Three results involving power products are, for real x:

(a)
$$\prod_{n=0}^{\infty}(1+x^{2^n}) = \frac{1}{1-x} \qquad (-1 < x < 1)$$

(b)
$$\prod_{n=1}^{\infty}(1-\frac{x^2}{n^2}) = \frac{\sin \pi x}{\pi x}$$

(c)
$$\prod_{n=1}^{\infty}(1-\frac{4x^2}{(2n-1)^2}) = \cos \pi x$$

EXAMPLE 19

Prove the power product (a): For $|x| < 1$, $\displaystyle\prod_{n=0}^{\infty}(1+x^{2^n}) = \frac{1}{1-x}$

SOLUTION

Note that $(1+x)(1+x^2) = 1 + x + x^2 + x^3$ and that $(1+x)(1+x^2)(1+x^4) = 1 + x + x^2 + x^3 + x^4 + x^5 + x^6 + x^7$. In general, if the identity is true for $n = 1$ and we assume it true for some $n \geq 1$, then by multiplying both sides by $1 + x^{2^{n+1}}$ (with $q = 1 + 2 + 2^2 + \cdots + 2^n = 2^{n+1} - 1$), we obtain

$$(1+x)(1+x^2)(1+x^4)\cdots(1+x^{2^n})(1+x^{2^{n+1}}) = (1+x+x^2+\cdots+x^q)(1+x^{2^{n+1}})$$

$$= (1+x+x^2+\cdots+x^q)+(1+x+x^2+\cdots+x^q)\cdot x^{2^{n+1}}$$

$$= 1+x+x^2+\cdots+x^q+x^{2^{n+1}}+x^{1+2^{n+1}}\cdots+x^{1+2+2^2+\cdots+2^n+2^{n+1}}$$

This proves (by mathematical induction) that for all integers $n \geq 0$,

$$(1+x)(1+x^2)(1+x^4)\cdots(1+x^{2^n}) = 1+x+x^2+x^3+\cdots+x^q$$

where $q = \sum_{k=0}^{n+1} 2^k$. By letting $n \to \infty$, the desired result follows from the geometric series.

The next example leads to a fundamental result for infinite products.

EXAMPLE 20

(a) Prove by mathematical induction that for all integers $n \geq 2$,

$$\left(1+\frac{1}{3}\right)\left(1+\frac{1}{8}\right)\left(1+\frac{1}{15}\right)\cdots\left(1+\frac{1}{n^2-1}\right) = \frac{2n}{n+1}$$

(b) Show that for all integers $n \geq 1$,

$$\left(1+\frac{1}{1}\right)\left(1+\frac{1}{2}\right)\left(1+\frac{1}{3}\right)\cdots\left(1+\frac{1}{n}\right) = n+1$$

(c) Using the results from (a) and (b), establish the following

$$\prod_{n=2}^{\infty}\left(1+\frac{1}{n^2-1}\right) = 2 \qquad \text{and} \qquad \prod_{n=1}^{\infty}\left(1+\frac{1}{n}\right) = \infty$$

SOLUTION

(a) The identity is true for $n = 2$ since $4/3 = 2\cdot2/(2+1)$. Assume true for some $n \geq 2$, and multiply both sides of the resulting identity by the next factor $1 + 1/[(n+1)^2 - 1]$. Thus:

$$\left(1+\frac{1}{3}\right)\left(1+\frac{1}{8}\right)\left(1+\frac{1}{15}\right)\cdots\left(1+\frac{1}{n^2-1}\right)\left(1+\frac{1}{(n+1)^2-1}\right) = \frac{2n}{n+1}\cdot\left(1+\frac{1}{(n+1)^2-1}\right)$$

$$= \frac{2n}{n+1}\cdot\frac{(n+1)^2}{n^2+2n} = \frac{2(n+1)}{n+2}$$

which is the identity for $n + 1$. Hence, by induction, the identity is true for all $n \geq 2$.

(b) Here we have

$$\left(1+\frac{1}{1}\right)\left(1+\frac{1}{2}\right)\left(1+\frac{1}{3}\right)\cdots\left(1+\frac{1}{n}\right) = 2\cdot\frac{3}{2}\cdot\frac{4}{3}\cdots\frac{n+1}{n} = n+1$$

(c) By definition of infinite products,

$$\prod_{n=2}^{\infty}\left(1+\frac{1}{n^2-1}\right) = \lim_{n\to\infty}\frac{2n}{n+1} = 2 \qquad \text{and} \qquad \prod_{n=1}^{\infty}\left(1+\frac{1}{n}\right) = \lim_{n\to\infty}(n+1) = \infty$$

The two infinite products of Example 19 suggest a relationship between $\prod(1+a_n)$ and $\sum a_n$. There we observed that $\prod_{n=2}^{\infty}\left(1+\frac{1}{n^2-1}\right)$ converges, while $\prod_{n=1}^{\infty}\left(1+\frac{1}{n}\right)$ diverges. On the other hand, the series $\sum\frac{1}{n^2-1}$ converges (but not to 2), while $\sum\frac{1}{n}$ diverges. Indeed, a basic theorem on infinite products is the following.

THEOREM 9.20

Suppose $a_n \geq 0$ for all sufficiently large n. Then the real infinite product $\prod(1 + a_n)$ converges iff the series $\sum a_n$ converges.

> *Proof:* Since the inequality $1 + x \leq e^x$ holds for all $x \geq 0$, note that $(1 + x)(1 + y) \leq e^x e^y = e^{x+y}$. Hence it follows that if $a_n \geq 0$,
>
> (*) $\qquad a_1 + a_2 + a_3 + \cdots < (1 + a_1)(1 + a_2)(1 + a_3)\cdots \leq e^{a_1 + a_2 + a_3 + \cdots}$
>
> If $\sum a_n$ converges, then $\prod(1 + a_n)$ is bounded. Since $1 + a_n \geq 1$ for $a_n \geq 0$, the partial products p_n are monotone increasing, and p_n is bounded above. Therefore $\{p_n\}$ converges to its least upper bound and $\prod(1 + a_n)$ converges. Conversely, suppose that $\prod(1 + a_n)$ converges. Then by (*) the series of positive terms $\sum a_n$ is bounded and therefore converges. ◇

COROLLARY

Suppose $0 \leq a_n < 1$ for all sufficiently large n. Then the infinite product $\prod(1 - a_n)$ converges iff the series $\sum a_n$ converges.

> *Proof:* In this case, the partial products are bounded below and are monotone decreasing, so $\{p_n\}$ converges to its greatest lower bound. Conversely, suppose that $\prod(1 - a_n)$ converges to a (nonzero) limit L. Then from (*), since all terms are positive (for large enough n) we obtain $1/L \geq 1/e^{-a_1 - a_2 - a_3 - \cdots} = e^{a_1 + a_2 + a_3 + \cdots}$ from some point on. Thus $\sum a_n$ is bounded and therefore converges. ◇

Theorem **9.20** does not imply in general that the product $\prod(1 + a_n)$ converges if the series $\sum a_n$ does so, (where the terms a_n are allowed to be negative). In that case we can invoke the general result for complex variables:

THEOREM 9.21 (Coriolis Test)

If both $\sum c_n$ and $\sum c_n^2$ converge for complex c_n, then the complex product $\prod(1 + c_n)$ converges.

This theorem was originally attributed to Cauchy. (The converse is not true; see Problem 43 for a simple counterexample.)

We mention two further significant results involving infinite products for complex variables. The first can be derived from the series expansion obtained in Example 18 (see Problems 45–47 for details). These two results justify the power products **(b)** and **(c)** above.

$$(9.22) \qquad \sin z = z\left(1 - \frac{z^2}{\pi^2}\right)\left(1 - \frac{z^2}{4\pi^2}\right)\left(1 - \frac{z^2}{9\pi^2}\right)\cdots = z\prod_{n=1}^{\infty}\left(1 - \frac{z^2}{n^2\pi^2}\right)$$

$$(9.23) \qquad \cos z = \left(1 - \frac{z^2}{(\pi/2)^2}\right)\left(1 - \frac{z^2}{(3\pi/2)^2}\right)\left(1 - \frac{z^2}{(5\pi/2)^2}\right)\cdots = \prod_{n=1}^{\infty}\left(1 - \frac{4z^2}{(2n-1)^2\pi^2}\right)$$

THEOREM 9.24 (Weierstrass Factor Theorem)

If $f(z)$ is an entire function having simple zeroes at $a_1, a_2, a_3, \cdots, a_n, \cdots$ such that $|a_1| < |a_2| < |a_3| \cdots$ and $\lim\limits_{n \to \infty} |a_n| = \infty$, then

$$f(z) = ce^{dz/c}\prod_{n=1}^{\infty}(1 - \frac{z}{a_n})e^{z/a_n} \qquad \text{where } c = f(0) \text{ and } d = f'(0)$$

PROBLEMS

Probleims 1–7 can be solved using the semicircle C_r of Examples 1 and 2 above. In each case, you are to evaluate the given integral by contour integration, showing all details.

1. $\displaystyle\int_0^\infty \frac{dx}{x^6+1}$

2. $\displaystyle\int_0^\infty \frac{x^2}{x^6+1}\,dx$

3. By making the appropriate adjustments to the details in the steps of Example 2, establish the general

formula (for $a, b > 0$) $\displaystyle\int_0^\infty \frac{\cos bx}{x^2+a^2}\,dx = \frac{\pi}{2a}e^{-ab}$.

4. $\displaystyle\int_0^\infty \frac{dx}{x^4+x^2+1}$

5. $\displaystyle\int_0^\infty \frac{dx}{(x^2+1)(x^2+4)}$

6. $\displaystyle\int_0^\infty \frac{(\ln x)^2}{x^2+1}\,dx$

7. Show that $\displaystyle\int_0^\infty \frac{x\sin x}{x^2+1}\,dx = \frac{\pi}{2e}$.

8. Taking C_r to be the rectangle having vertices $A(-r)$, $B(r)$, $D(r+\pi i)$, $E(-r+\pi i)$, prove $\displaystyle\int_0^\infty \frac{\cosh\frac{2}{3}x}{\cosh x}\,dx = \pi$.
   [**Hint:** $\cosh z = \cosh x \cos y + i \sinh x \sin y$.]

9. Taking C_r to be the boundary of the wedge-shaped region shown in the figure, with vertices $A(0)$, $B(r)$, and $D(re^{\pi i/4})$, show that
$$\int_0^\infty \sin x^2\,dx = \int_0^\infty \cos x^2\,dx = \frac{\sqrt{2\pi}}{4}$$
   [**Hint:** First evaluate the integral $\displaystyle\int_{C_r} e^{iz^2}\,dz$ ($\sin 2t > t$ for $0 \le t \le \pi/4$). You will find it necessary to use the statistics integral $\int e^{-x^2}dx = \sqrt{\pi}$.]

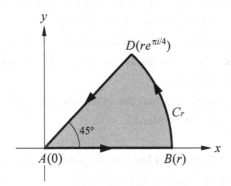

10. Describe the geometric effects of the following mappings given in terms of complex variables and give their components u and v where $w = u + iv$.
    (a) $w = 3z$
    (b) $w = iz$
    (c) $w = (3 + 4i)z$

11. Consider the linear transformation given by

$$\begin{cases} u = 3x - 4y \\ v = 4x + 3y \end{cases}$$

(a) Find the images of the perpendicular lines $y = x + 3$ and $y = -x + 1$.

(b) Characterize this mapping as a complex function of the form $w = cz$ by finding c.

(c) This transformation rotates all figures through some fixed angle θ. Find θ.

12. Repeat the analysis of Problem 11 for the transformation

$$\begin{cases} u = 4x - 3y \\ v = -3x - 4y \end{cases}$$

13. Consider the complex inversion $w = z^{-1}$.

(a) Plot the points $c = \tfrac{4}{5} - \tfrac{3}{5}i$, $2 + 2i$, and $-\tfrac{1}{2}i$ and their images calculated from $w = z^{-1}$ on the same set of axes. Does this result show the basic properties of a complex inversion?

(b) Show by direct substitution that the vertical line $x = 2$ maps to the circle $|z - \tfrac{1}{4}| = \tfrac{1}{4}$.

14. **The fractional linear map**

Consider the general mapping (for complex constants a, b, c, and d)

$$w = \frac{ax + b}{cz + d} \qquad (ad \neq bc)$$

Show that any such mapping (with $c \neq 0$) is the product of a complex linear mapping, a complex inversion, and another complex linear map, by performing the necessary algebra on the successive equations (for appropriate values for D and C you are to find)

$$w = Dw_2 + C, \qquad w_2 = \frac{1}{w_1}, \qquad \text{and} \qquad w_1 = cz + d$$

15. Using the sequence of maps from Problem 14, establish the final effect of the fractional linear mapping given by $w = \dfrac{3z + 2}{z + 1}$ for each of the following geometric objects:

(a) The point $c = i$.

(b) The circle $|z + 2| = 1$.

(c) The line $y = -\tfrac{1}{2}$ (parametric form $z = t - \tfrac{1}{2}i$, t real).

16. Show that the squaring map takes the upper half-plane and non-negative x-axis one-to-one to the entire uv-plane. [**Hint:** For one-to-oneness, show that each ray T from the origin to infinity in the xy-plane is mapped one-to-one to a ray T' from the origin to infinity in the uv-plane.]

17. Show that the squaring map on the entire complex plane takes the circle $|z - a| = a$ (real a) onto the cardioid $r = 2a^2(1 + \cos\theta)$. [Polar coordinates for the uv-plane are defined by $r = (u^2 + v^2)^{1/2}$, $\theta = \tan^{-1}(v/u)$. In this problem, the squaring map is not one-to-one.]

18. Prove that if $f(z) = u(x, y) + iv(x, y)$ is analytic, then the system of curves $u(x, y) = \alpha$, $v(x, y) = \beta$ is a mutually orthogonal family of curves in the xy-plane.

19. **The sine mapping $w = \sin z$**

Show that the mapping $w = \sin z$ takes horizontal lines to ellipses, and find the foci of each of these ellipses. Are the system of ellipses and hyperbolas (page 162) confocal?

20. **The sine mapping $w = \sin z$ and one-to-oneness**

Prove that $w = \sin z$ maps the vertical strip described above one-to-one onto the entire uv-plane. [**Hint:** Show that each horizontal line segment $y = b \neq 0$, $-\pi/2 < x < \pi/2$ in the xy-plane is mapped one-to-one to a half-ellipse in the uv-plane, with distinct line segments mapping to distinct half-ellipses. (Examine also the points on the x-axis.)]

21. The exponential mapping $w = \exp z$

(a) Verify that $w = e^z$ represents the transformation

$$\begin{cases} u = e^x \cos y \\ v = e^x \sin y \end{cases}$$

and that the coordinate grid in the xy-plane maps to a mutually orthogonal system of concentric circles and lines through their common center in the uv-plane, as illustrated in the figure.

(b) Show that the exponential mapping maps the infinite horizontal strip, $-\pi \le y \le \pi$, one-to-one onto the entire uv-plane. [**Hint:** Show that $e^x \cos y = e^a \cos b$, $e^x \sin y = e^a \sin b$ has a unique solution in the infinite strip $-\pi < y \le \pi$.]

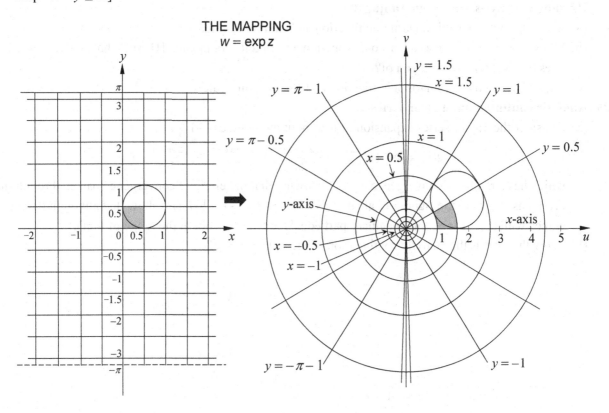

THE MAPPING
$w = \exp z$

22. The logarithm mapping $w = \ln z$

Show that $w = \ln z$ maps the annular ring $a \le |z| < b$ onto the rectangle bounded by the lines $u = \ln a$, $u = \ln b$, and $v = \pm \pi$. What is the image of the annular sector defined by $a \le r \le b, p \le \theta \le q$? Is this mapping one-to-one? [**Hint:** Don't forget that $\ln z$ is the inverse of e^z over the region defined in Problem 21.]

23. Dynamic systems

In Example 8, $g(z)$ was defined as the second iterate of $f(z) = z^2 + 0.16$, or $g(z) = (z^2 + 0.16)^2 + 0.16$. Two fixed points of $g(z)$ are the two fixed points of $f(z)$, which are the roots of $f(z) - z = z^2 - z + 0.16 = 0$. [Verify this fact in general for functions like $g(z)$.] Find the other factor (besides $z^2 - z + 0.16$) of $z^4 + 0.32z^2 - z + 0.185$ as mentioned in Example 7 by performing the long division calculation as indicated below:

$$z^2 - z + 0.16\overline{)z^4 + 0.32z^2 - z + 0.1856}$$

24. Dynamic systems and Julia sets

(a) Show that if $f(z) = z^2 - 2$ then $f^2(z) \equiv g(z) = z^4 - 4z^2 + 2$. Find the four fixed points of g by using the techniques of Example 8 and Problem 23, then classify them as either attractive or repelling fixed points.

(b) Which of these four points belong to J_{-2}?

25. The Mandelbrott set: proof of (9.7) and (9.8)

(a) Prove **(9.7)** using the triangle inequality.

(b) Prove **(9.8)**.

26. The Mandelbrott set: intersection with real axis

Show that the closed interval of real numbers $[-2, \frac{1}{4}]$ is a subset of M, but any real number outside this interval does not belong to M. [**Hint:** For the first part, show that if $-2 \leq c \leq \frac{1}{4}$ then, for $c \geq 0$, $0 \leq c_n \leq \frac{1}{2}$ for all n; for $c < 0$, show that if $-2 \leq c < 0$ (thus $c = -2 + \varepsilon$ for $0 \leq \varepsilon < 2$), then $-2 < c_n < 2 - \varepsilon$ for all n.]

27. Harmonic analysis

Show that $\ln(x^2 + y^2)$, $x \neq 0$, is a solution of LaPlace's partial differential equation, and use Theorem **9.11** to show that $\tan^{-1}(y/x)$ is a harmonic conjugate of $\ln(x^2 + y^2)^{1/2} = \frac{1}{2}\ln(x^2 + y^2)$.

28. Harmonic analysis: harmonic conjugates

(a) Show that $x/(x^2 + y^2)$ is a harmonic function and find its conjugate.

(b) Show that $xy/(x^2 + y^2)^2$ is a harmonic function and find its conjugate. [**Hint:** What analytic function does $(u + i \cdot 2xy)^{-1}$ remind you of?]

(c) Show that $e^{-y}\cos x$ is a harmonic function and find its conjugate.

29. Analytic continuation: Taylor series

(a) Consider the Taylor series expansion of $\ln z$ about the point $c = -1 + i$

$$g(z) \equiv \ln c + \frac{z-c}{c} - \frac{(z-c)^2}{2c^2} + \frac{(z-c)^3}{3c^3} - \frac{(z-c)^4}{4c^4} + \cdots$$

which has a radius of convergence $r = \sqrt{2}$, showing that the circle of convergence passes through the origin, as shown in the figure. Evaluate this series at $z_1 = -1 - \frac{1}{3}i$ and show that although the real part agrees with that of $\ln z_1$, the imaginary part equals $\pi + \tan^{-1}2$ (not in the range of $\ln z$!).

(b) Show that $g(z)$ is the analytic continuation of $\ln z$ to $|z + c| < \sqrt{2}$.

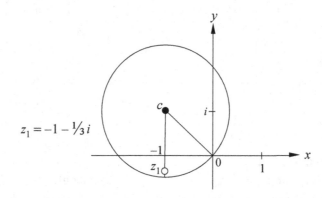

30. Analytic continuation: proving identities

Use the corollary of Theorem **9.12** to prove the identities for all complex z and w

(a) $\sin(z + w) = \sin z \cos w + \cos z \sin w$

(b) $\cos 2z = \cos^2 z - \sin^2 z$.

31. Analytic continuation: proving identities

(a) Use the corollary of Theorem **9.12** to prove that for z in a certain region R, $\ln \sqrt{z} = \frac{1}{2}\ln z$.

(b) The corresponding identity for real numbers is $\ln\sqrt{x} = \frac{1}{2}\ln x$, valid for $x > 0$. For what complex numbers z is the identity in (a) valid?

32. Show that if $\pi < \theta < 3\pi$ then $\sqrt{r}e^{i\theta/2} = -\sqrt{z}$ where $\sqrt{z}$ is the square root radical as defined in Chapter 2.

33. Analytic continuation: square root function

Define the analytic function $\sqrt{r}\,e^{i\theta} = f_k(z)$ for $k = 1, 2, 3, 4$ on the respective domains

$$R_1: 0 < \theta < \pi, \qquad R_2: \frac{\pi}{2} < \theta < \frac{3\pi}{2}, \qquad R_3: \frac{3\pi}{2} < \theta < 2\pi, \qquad R_4: \frac{3\pi}{2} < \theta < \frac{5\pi}{2}$$

Verify that $f_1(z)$ is an analytic continuation of $f_2(z)$, that $f_3(z)$ is that of $f_2(z)$, and $f_4(z)$ that of $f_3(z)$, and show that $f_4(z) \neq f_1(z)$ for any point in $R_1 \cap R_2$.

34. Rouche's theorem

Show that the five solutions of the equation $z^5 + 8z^4 + 4z + 1 = 0$ all lie in the closed disk $|z| \le 1$.

35. Theorem: The infinite product $(1 - \frac{1}{2})(1 + \frac{1}{3})(1 - \frac{1}{4})(1 + \frac{1}{5})\cdots$ converges to $\frac{1}{2}$.

(a) Give an Euler-style argument.

(b) Give a rigorous proof using limits and the mathematical definition of an infinite product.

36. An Euler-style argument indicates that $\frac{2}{3} \cdot \frac{6}{9} \cdot \frac{18}{27} \cdot \frac{54}{81} \cdots = 2 \cdot 2 \cdot 2 \cdot 2 \cdots = \infty$. What value does the definition for infinite products assign to this product?

37. Verify the infinite product given above: $\dfrac{3}{4} \cdot \dfrac{8}{9} \cdot \dfrac{15}{16} \cdots \dfrac{n^2 - 1}{n^2} \cdots = \dfrac{1}{2}$ $(n \ge 2)$.

38. Show that the values

$$\frac{8}{9} \cdot \frac{24}{25} \cdot \frac{48}{49} \cdots \frac{(2n+1)^2 - 1}{(2n+1)^2} \cdots = \frac{\pi}{4} \quad \text{and} \quad \frac{8}{9} \cdot \frac{80}{81} \cdot \frac{224}{225} \cdots \frac{(6n-3)^2 - 1}{(6n-3)^2} \cdots = \sqrt{3}/2$$

can each be derived from one of the above power products (a), (b), or (c).

39. Show that the infinite product $\dfrac{4}{3} \cdot \dfrac{36}{35} \cdot \dfrac{100}{99} \cdots \dfrac{(4n-2)^2}{(4n-2)^2 - 1} \cdots = \sqrt{2}$ can be derived from the power product **(c)**.

40. The product of the even integers divided by the product of the odd integers, although indetermimant, can be expressed as an infinite product: $\dfrac{2}{1} \cdot \dfrac{4}{3} \cdot \dfrac{6}{5} \cdots \dfrac{2n}{2n-1} \cdots$ Is this infinite product convergent or divergemt?

(Use Theorem 9.20 to answer.)

41. The partial products p_n of the infinite product defined in Problem 40 are asymptotic to $\sqrt{\pi n}$ (that is, $p_n/\sqrt{\pi n}$ converges to 1 as $n \to \infty$.). Show this. (Thus for large n, $p_n \approx \sqrt{\pi n}$.) [**Hint:** Use the result of Problem 38 on the infinite product $\dfrac{2 \cdot 4}{3 \cdot 3} \cdot \dfrac{4 \cdot 6}{5 \cdot 5} \cdot \dfrac{6 \cdot 8}{7 \cdot 7} \cdots.]$

42. Show that the power products **(b)** and **(c)** are special cases of **(9.22)** and **(9.23)** (see Problems 45–47).

43. Discuss the product indicated by

$$\left(1 - \frac{1}{\sqrt[4]{2}}\right)\left(1 + \frac{1}{\sqrt[4]{2}}\right)\left(1 + \frac{1}{\sqrt{2}}\right)\left(1 + \frac{1}{2}\right) \cdot \left(1 - \frac{1}{\sqrt[4]{3}}\right)\left(1 + \frac{1}{\sqrt[4]{3}}\right)\left(1 + \frac{1}{\sqrt{3}}\right)\left(1 + \frac{1}{3}\right)\cdots$$

and show it is a counterexample for the converse of Coriolis' theorem.

44. Prove that if $(1 + x)(1 + x^3)(1 + x^5)(1 + x^7)\cdots$ converges, then so does $(1 + x)(1 + x^2)(1 + x^3)(1 + x^4)\cdots$.

45. Complete the following details establishing the product formula **(9.22)** from **(9.19)**:

(1) Begin with the formula **(9.19)** from Example 18:

$$\cot z - \frac{1}{z} = \sum_{n=1}^{\infty} \frac{2z}{z^2 - n^2\pi^2}$$

(2) Integrate both sides from real $\varepsilon > 0$ to complex w:

$$\int_{\varepsilon}^{w}\left(\cot z - \frac{1}{z}\right)dz = \sum_{n=1}^{\infty}\int_{\varepsilon}^{w}\left(\frac{2z}{z^2 - n^2\pi^2}\right)dz$$

Hence,

$$\ln\frac{\sin z}{z}\bigg|_{\varepsilon}^{w} = \sum_{n=1}^{\infty}\left[\ln\left(z^2 - n^2\pi^2\right)\right]\bigg|_{\varepsilon}^{w}$$

(3) Substitute the limits of integration into these functions; the result is (after changing the variable w to z):

$$\ln\frac{\sin z}{z} - \ln\frac{\sin \varepsilon}{\varepsilon} = \sum_{n=1}^{\infty} \ln\left(\frac{z^2 - n^2\pi^2}{\varepsilon^2 - n^2\pi^2}\right)$$

(4) Take the limit as $\varepsilon \to 0$ to obtain **(9.22)**

$$\ln\frac{\sin z}{z} = \sum_{n=1}^{\infty} \ln\left(\frac{z^2 - n^2\pi^2}{-n^2\pi^2}\right)$$

46. Using Example 18 as model, show that Mittag-Leffler's theorem leads to the series expansion

(9.25) $$\tan z = \sum_{n=1}^{\infty} \frac{8z}{(2n-1)^2\pi^2 - 4z^2}$$

47. As in Problem 45 in deriving **(9.22)** from **(9.19)**, use **(9.25)** to obtain **(9.23)**.

APPENDIX A:
ANSWERS TO SELECTED PROBLEMS

CHAPTER 1

1. $4 \pm 5i$.

5. (a) $(3 - 4i)^{-1} = \begin{bmatrix} 3 & -4 \\ 4 & 3 \end{bmatrix}^{-1} = \dfrac{1}{25}\begin{bmatrix} 3 & 4 \\ -4 & 3 \end{bmatrix}$

(b) $(2 + 3i)(3 + 4i)/25 = (6 - 12 + 17i)/25$
$= -\%_{25} + {}^{17}\!\%_{25}i$

(c) $\dfrac{2+3i}{3-4i} = \dfrac{2+3i}{3-4i} \cdot \dfrac{3+4i}{3+4i} = \dfrac{-7+17i}{25}$

6. $\begin{bmatrix} a & b \\ -b & a \end{bmatrix}\begin{bmatrix} c & d \\ -d & c \end{bmatrix} = \begin{bmatrix} ac-bd & ad+bc \\ -bc-ad & -bd+ac \end{bmatrix}$
$= \begin{bmatrix} c & d \\ -d & c \end{bmatrix}\begin{bmatrix} a & b \\ -b & a \end{bmatrix}$

8. $a + bi = a\mathrm{I} + (b\mathrm{I})\mathrm{J} = a\mathrm{I} + b\mathrm{J} =$
$= \begin{bmatrix} a & 0 \\ 0 & a \end{bmatrix} + \begin{bmatrix} 0 & b \\ -b & 0 \end{bmatrix} = \begin{bmatrix} a & b \\ -b & a \end{bmatrix}$

10. (a) $\sqrt{2}, \sqrt{8}$; **(b)** $\sqrt{2}\operatorname{cis}3\pi/4, \sqrt{8}\operatorname{cis}3\pi/4$;
(c) $\operatorname{Arg}zw = \operatorname{Arg}(-1 + i)(2 + 2i) = \operatorname{Arg}(-4) = \pi$, $\operatorname{Arg}z + \operatorname{Arg}w = 3\pi/4 + \pi/4 = \pi$

11. (2) $\overline{(x + iy)(u + iv)} = \overline{(xu - yv) + i(xv + yu)} = (xu - yv) - i(xv + yu)$, while $\bar{z}\cdot\bar{w} = (x - iy)(u - iv) = (xu - yv) - i(xv + yu)$

13. $2 + 2i = \sqrt{8}\operatorname{cis}45°$, $s\operatorname{cis}\varphi$ where $s = \sqrt{\left(\dfrac{\sqrt{3}+1}{2}\right)^2 + \left(\dfrac{\sqrt{3}-1}{2}\right)^2} = \sqrt{2}$ and $\varphi = \tan^{-1}\dfrac{\sqrt{3}-1}{\sqrt{3}+1} = \tan^{-1}(2 - \sqrt{3}) = \tan^{-1}15°$

Product $= (2 + 2i)\left(\dfrac{\sqrt{3}+1}{2} + \dfrac{\sqrt{3}-1}{2}i\right) =$
$2 + 2\sqrt{3}i = \sqrt{2^2 + (2\sqrt{3})^2}\operatorname{cis}\left(\tan^{-1}\dfrac{2\sqrt{3}}{2}\right)$

$= \sqrt{16}\operatorname{cis}(\tan^{-1}\sqrt{3}) = 4\operatorname{cis}60°$
In polar form, product $= \sqrt{8}\operatorname{cis}45°\cdot\sqrt{2}\operatorname{cis}15° = \sqrt{8}\cdot\sqrt{2}\operatorname{cis}(45° + 15°) = 4\operatorname{cis}60°$

15. (1) $[(\sqrt{3} + 1)/2 + (\sqrt{3} - 1)i/2]^6 = [\sqrt{2}\operatorname{cis}15°]^6 = 8\operatorname{cis}90° = 8i$

(2) $\left(\dfrac{\sqrt{3}+1}{2}\right)^2 + 2\left(\dfrac{\sqrt{3}+1}{2}\cdot\dfrac{\sqrt{3}-1}{2}\right)i + \left(\dfrac{\sqrt{3}+1}{2}\right)^2 i^2$
$= \sqrt{3} + i;$
$(\sqrt{3} + i)^3 = (\sqrt{3})^3 + 3(\sqrt{3})^2i + 3\sqrt{3}i^2 + i^3 = 3\sqrt{3} - 3\sqrt{3} + i(9 - 1) = 8i.$

17. $2^{50}i$

19. $(\operatorname{cis}\theta)^{-1} = \dfrac{1}{\cos\theta + i\sin\theta} = \dfrac{\cos\theta - i\sin\theta}{\cos^2\theta + \sin^2\theta} = \cos(-\theta) + i\sin(-\theta)$. Let $n = -m$ ($m > 0$) be an integer. Then $(\operatorname{cis}\theta)^n = (\operatorname{cis}\theta)^{-m} = [\cos(-\theta) + i\sin(-\theta)]^m = \cos(-m\theta) + i\sin(-m\theta) = \operatorname{cis}n\theta.$

21. $(\operatorname{cis}2k\pi/n)^n = \operatorname{cis}(n\cdot2k\pi/n) = \operatorname{cis}2k\pi = 1.$

23. By Problem 22, 5th roots of $32i = 32\operatorname{cis}90°$ are $32^{1/5}\operatorname{cis}(90°/5 + 360°/5) = 2\operatorname{cis}(18° + 72k°)$ for $k = 0, 1, 2, 3, 4$. These are the vertices of a regular pentagon inscribed on the circle $|z| = 2$.

25. By definition of polar coordinates, $\arg c = \operatorname{Arg}c + 2k\pi$ (k an integer). If $k \neq 0$, then either $\arg c > -\pi + 2\pi = \pi$ or $\arg c \leq \pi - 2\pi = -\pi$. By (1.3), $\operatorname{Arg}z + \operatorname{Arg}w = \operatorname{arc}z + \operatorname{arg}w = \operatorname{arg}zw$. If $\operatorname{arg}zw \neq \operatorname{Arg}zw$, then $\operatorname{arg}zw > \pi$ or $\operatorname{arg}zw \leq -\pi$. But this contradicts the hypothesis.

CHAPTER 2

3. $4 - i$, $\sqrt{4-i} = \sqrt{\dfrac{\sqrt{17}+4}{2}} - i\sqrt{\dfrac{\sqrt{17}-4}{2}} \approx$

    $2.01533 - 0.24810i$.

5. $2 + 3i$, $1 - i$.

10. (b) $-i\sinh 2$.

15. $\cos z \cos w - \sin z \sin w = \tfrac{1}{4}(e^{iz}+e^{-iz})\cdot$

    $(e^{iw}+e^{-iw}) - \tfrac{1}{4i^2}(e^{iz}-e^{-iz})(e^{iw}-e^{-iw})$

    $= \tfrac{1}{4}(e^{iz+iw}+e^{iz-iw}+e^{-iz+iw}+e^{-iz-iw}+$

    $e^{iz+iw}-e^{iz-iw}-e^{-iz+iw}+e^{-iz-iw}) =$

    $\tfrac{1}{4}(2e^{iz+iw}+2e^{-iz-iw}) = \cos(z+w)$

16. $e^{2z} = 1 \Rightarrow e^{2x}\cos 2y = 1$, $e^{2x}\sin 2y = 0 \Rightarrow$
    $e^{4x}(\cos^2 2y + \sin^2 2y) = 1^2 + 0^2$ or $e^{4x} = 1 \Rightarrow$
    $x = 0 \Rightarrow \sin 2y = 0 \Rightarrow 2y = n\pi$ (n = integer),
    $\cos 2y = 1 \Rightarrow n$ even $\Rightarrow y = k\pi$ (k = integer) $\Rightarrow$
    $z = k\pi i$.

19. $\cos(\ln 3) + i\sin(\ln 3)$

23. [Region (c)] By using the set-up of Example 9,
    $|u\,\mathrm{Arg}z + v\ln|z|| \le |u||\mathrm{Arg}z| + |v||\ln|z|| <$
    $2\cdot\pi/4 + 2\ln 2 \approx 2.9571 < \pi$.

CHAPTER 3

1. (b) $\ln(z^2 + z + 1) + \dfrac{2z^2 + z}{z^2 + z + 1}$

5. $\dfrac{d}{dz}e^w = e^w\dfrac{dw}{dz} = 1 \Rightarrow \dfrac{dw}{dz} = \dfrac{1}{e^w} = \dfrac{1}{z}$.

9. $\displaystyle\int e^z z^3 dz = e^z z^3 - \int e^z\cdot 3z^2 dz = e^z z^3 - e^z\cdot 3z^2 +$

    $\displaystyle\int e^z\cdot 6z\,dz = e^z z^3 - e^z\cdot 3z^2 + e^z\cdot 6z - \int e^z\cdot 6\,dz$

    $= e^z(z^3 - 3z^2 + 6z - 6) + C$

11. $u = x$, $v = -y \Rightarrow \dfrac{\partial u}{\partial x} = 1$, $\dfrac{\partial u}{\partial y} = 0$, $\dfrac{\partial v}{\partial x} = 0$,

    $\dfrac{\partial v}{\partial y} = -1 \Rightarrow \dfrac{\partial u}{\partial x} \ne \dfrac{\partial v}{\partial y}$.

13. From the hint, $u(x + h, y + k) - u(x, y) =$
    $[u(x + h, y + k) - u(x, y + k)]$
        $+ [u(x, y + k) - u(x, y)]$

    $= [u_x(x, y + k)h + \varepsilon'h] + [u_y(x, y)k + \delta k]$
    where ε', $\delta \to 0$ as h, $k \to 0$. Make the substitution $\varepsilon' = \varepsilon - u_x(x, y + k) + u_x(x, y)$. By continuity of u_x, $\varepsilon \to 0$ as h, $k \to 0$, and **(3.17)** follows.

15. $\delta = \dfrac{\delta_1 h_1 + \delta_2 h_2}{h} = \delta_1\dfrac{h_1}{h} + \delta_2\dfrac{h_2}{h} \Rightarrow |\delta| =$

    $\left|\delta_1\dfrac{h_1}{h} + \delta_2\dfrac{h_2}{h}\right| \le |\delta_1|\dfrac{|h_1|}{|h|} + |\delta_2|\dfrac{|h_2|}{|h|} \le |\delta_1| + |\delta_2|$

    Since ε_1 and $\varepsilon'_1 \to 0$ as $h \to 0$, then $\delta_1 \to 0$.
    Similarly, $\delta_2 \to 0$, and it follows that $\delta \to 0$.

23. (a) 1; **(b)** $\tfrac{1}{6}$.

27. $u = x^2 - y^2 + x + c$ (where c is an arbitrary real).

CHAPTER 4

1. $\displaystyle\int_C f(z)dz = \int_0^1 (4t + 2ti)(1 + i)dt$

    $= \displaystyle\int_0^1 [4t - 2t + i(4t + 2t)]dt = \int_0^1 [2t + 6ti]dt$

    $= 1 + 3i$

3. 1

5. (b) $^7/_{12} + {^{16i}/_{21}}$

7. $72 - 9i$

9. -2

11. $(\sinh 1 + \sinh 2)i$

15. $2\pi i$

17. By the constant multiple property of limits,

    $\displaystyle\int_C cf(z)dz = \lim_{n\to\infty}\sum_{k=1}^{\infty} cf(z_k{}^*)\Delta z_k$

    $= c\lim_{n\to\infty}\sum_{k=1}^{\infty} f(z_k{}^*)\Delta z_k = c\displaystyle\int_C f(z)dz$

21. 6.7×10^7 $[|\sin z| = (\sin^2 6t \cosh^2 t^2 + \cos^2 6t \cdot$
$\sinh^2 t^2)^{\frac{1}{2}} < \cosh t^2 < \cosh 16.]$

22. $(10^{3/2} - 1)/54 \approx 0.56709$

CHAPTER 5

1. (a) 0; **(b)** $4\pi i$

3. $\dfrac{2\pi i}{\sqrt{e}}$

5. (a) 0; **(b)** $-2\pi i$

8. $\displaystyle\int_0^{2\pi} \frac{1 + \frac{1}{2}e^{ti}}{\frac{1}{4}e^{2ti}} \cdot \frac{1}{2} ie^{ti} dt = 2\int_0^{2\pi}(\sin t + i\cos t)dt + i\int_0^{2\pi} dt.$

10. $\pi i/4$

11. πi

12. $\pi a/16$

18. $\frac{1}{4}$

21. $-\pi i/360$

23. $z = -a + ie^{it}$, $0 \le t \le 2\pi$, is the unit circle centered at $-a$, which excludes the origin. Hence

$$\frac{2\pi i}{-a} = \int_C \frac{dz}{z(z+a)} = \int_0^{2\pi} \frac{-e^{it}}{(-a + ie^{it})ie^{it}} dt$$
$$= \frac{1}{i}\int_0^{2\pi} \frac{dt}{a - ie^{it}} = \frac{1}{i}\int_0^{2\pi} \frac{dt}{a + \sin t - i\cos t}$$

$$= -i\int_0^{2\pi} \frac{a + \sin t + i\cos t}{(a + \sin t)^2 + \cos^2 t} dt$$

$$= -i\int_0^{2\pi} \frac{a + \sin x}{a^2 + 2a\sin x + 1} dx$$

$$+ \int_0^{2\pi} \frac{\cos x}{a^2 + 2a\sin x + 1} dx$$

Therefore, $\displaystyle\int_0^{2\pi} \frac{a + \sin t}{a^2 + 2a\sin t + 1} dt = \frac{2\pi}{a}$

25. 12

27. By using Liouville's theorem, since $\sin z$ is not constant $\sin z$ is unbounded. By using the definition, $\sin(x + iy) = \sin x \cosh y + i\cos x \sinh y = \cosh y$ (unbounded if $x = \pi/2$).

29. $\sinh^2 1$ at $z = i$. $[|\sin z| = (\sin^2 x + \sinh^2 y)^{\frac{1}{2}} \ge |\sinh y|.]$

CHAPTER 6

1. i

3. 0 $(|\frac{1}{2} + \frac{2}{3}i| = \frac{5}{6})$

5. Does not exist

7. 1

9. 201st term

11. Graph: Infinite polygon having sides of unit length whose vertices lie on a fixed circle passing through 0,1, and $\frac{8}{5} + \frac{4}{5}i$.

13. (a) $3, 3 + 3i$
  (b) $100, 100 + 100i$

15. (a) Hint for line BF: $F(1, 1)$ and $B(\tau, 0)$.
  (b) $G(b/\tau^2, 0)$, $J(b/\tau^3, \pi/2)$, $K(b/\tau^4, \pi)$, and $M(b/\tau^5, 3\pi/2)$.

18. $\sum i^n/n = \sum(-1)^n/2n + i\sum(-1)^{n+1}/(2n-1)$.

19. $\left| \dfrac{i^{n+1}}{(1+i)^{n+1}} \cdot \dfrac{(1+i)^n}{i^n} \right| = \left| \dfrac{i}{1+i} \right| = \dfrac{1}{\sqrt{2}} < 1$

21. If $\mathscr{I}(z) = 0$ then $\sum(\sin nz)/n = \sum(\sin nx)/n$ which converges. If $\mathscr{I}(z) \ne 0$, then with $\sin z = \sin nx\cosh ny + i\cos ny\sinh ny$, the absolute value of the nth term of the series equals
$$\frac{\sqrt{\cosh^2 ny - \cos^2 nx}}{n} > \frac{|\cosh ny| - 1}{n} \to \infty$$
as $n \to \infty$. By Theorem 6.3, the series diverges.

22. With $\frac{3}{5} + \frac{4}{5}i = e^{i\theta}$ ($\theta = \tan^{-1}\frac{4}{3}$), the series $= \sum \cos n\theta/n + i\sum \sin n\theta/n$ (convergent). For absolute convergence, $\sum |\frac{3}{5} + \frac{4}{5}i|^n/n = \sum 1/n$ (divergent).

23. $\ln \dfrac{\sqrt{5}}{2} + i\tan^{-1} 2$

25. Converges by the comparison test.

26. $\sqrt{2}$.

27. 3

28. No: series diverges for $|z| = 3$. Function valid for $z \neq 3$.

30. All z such that $|z + 2| \leq 4$.

36. $|U_n| \leq \left| \dfrac{1}{2\pi i} \right| \displaystyle\int_C \left| \dfrac{f(w)W^{n+1}}{w - z} \right| |dw|$

$\leq \dfrac{B}{2\pi r}(s/r)^{n+1} l(C) = Bs(s/r)^n \to 0$

40. $g(z) = \dfrac{z}{(1-z)^2}$

45. If $z \neq 0$ $\sin z/z^2$ is the quotient of two analytic functions in the region $|z| > 0$, so is analytic. But at $z = 0$, since $\lim \sin z/z = 1$ as $z \to 0$, $\lim \sin z/z^2$ does not exist and the singularity cannot be removed.

47. Conjecture is true.

53. If p_n represents the number of plus signs in the sum of the first n terms, and q_n the number of minus signs, then $s_n = \varepsilon_1 + \varepsilon_2 + \varepsilon_3 + \cdots + \varepsilon_n = p_n - q_n$. But $p_n + q_n = n$ so $s_n = p_n - (n - p_n) = 2p_n - n < 2(n/2 + C) - n = 2C$.

CHAPTER 7

1. Uniform, on any bounded region.

2. (a) No; **(b)** Uniform on any closed half-plane $x \geq r > 0$.

3. (a) $\dfrac{1}{n}\sqrt{1 + n^2 z} = \sqrt{\dfrac{1}{n^2} + z} \to \sqrt{z}$

5. (a) z^3; **(c)** $\displaystyle\int_0^z nw\sin\dfrac{w^2}{n}dw = \dfrac{n^2}{2}(1 - \cos\dfrac{z^2}{n})$

$= \dfrac{1 - \cos\theta}{\theta/2} \cdot \dfrac{z^4}{4} \to \dfrac{z^4}{4}$ where $\theta = \dfrac{z^2}{n}$.

9. Accuracy guaranteed to only one decimal place when rounded.

11. With z replaced by i, the series converges to

$-\dfrac{1}{1\cdot 2} + \dfrac{1}{3\cdot 4} - \dfrac{1}{5\cdot 6} + \cdots$

$+ i\left(\dfrac{1}{2\cdot 3} - \dfrac{1}{4\cdot 5} + \dfrac{1}{6\cdot 7} + \cdots\right)$

each real series converging by the theorem on alternating series. Since $(z + 1)\ln(z + 1) - z$ is continuous at $z = i$, taking imaginary parts leads to the result

$\dfrac{1}{2\cdot 3} - \dfrac{1}{4\cdot 5} + \dfrac{1}{6\cdot 7} + \cdots$

$= \mathscr{I}[(z + 1)\ln(z + 1) - z] = \ln\sqrt{2} + \pi/4 - 1.$

CHAPTER 8

1. $\dfrac{-3z^2 - z + 3}{z^2(1 + z)}$

3. $-1/[z(z + 2)(z + 1)^2]$

5. $\dfrac{1}{(z - 1)^3} - \dfrac{1}{2(z - 1)^2} + \dfrac{1}{3(z - 1)}$

$-\dfrac{1}{4} + \dfrac{z - 1}{5} - \dfrac{(z - 1)^2}{6} + \cdots$

7. (a) $a_{-1} = \dfrac{1}{1!}\displaystyle\lim_{z \to 0}\left(\dfrac{-3z^2 - z + 3}{1 + z}\right)'$

$= \displaystyle\lim_{z \to 0}\dfrac{(-6z - 1)(1 + z) + 3z^2 + z - 3}{(1 + z)^2} = -4$

(b) $a_{-1} = \dfrac{1}{2!}\displaystyle\lim_{z \to 1}\left(\dfrac{\ln z}{z - 1}\right)''$

$= \dfrac{1}{2}\displaystyle\lim_{z \to 1}\dfrac{-1/z^2 + 4/z - 3 + 2\ln z}{(1 - z)^3} = \dfrac{1}{3}$

9. $(9z^2 - 7z + 1)/[3z^2(1 - z)]$

10. (b) -1 at $z = 0$, $^3/_2$ at $z = 1$, $-^1/_2$ at $z = -1$

(c) $-2\pi i$.

13. $\dfrac{2\pi}{5}$

15. $\dfrac{5\pi}{32}$

17. $\dfrac{z^{-3}}{3!} - \dfrac{z^{-1}}{5!} + \dfrac{z}{7!} - \dfrac{z^3}{9!} + \dfrac{z^5}{11!} - \cdots$

19. Using series, $a_{-1} = \dfrac{1}{2!}\displaystyle\lim_{z \to 0}\left(z^3\dfrac{\csc z}{z^2}\right)^{(2)}$

$= \dfrac{1}{2}\displaystyle\lim_{z \to 0}\left(\dfrac{z}{\sin z}\right)'' =$

$$= \frac{1}{2} \lim_{z \to 0} \left[\frac{1}{1 - \dfrac{z^2}{6} + \dfrac{z^4}{120} - \dfrac{z^6}{5040} + \cdots} \right]''$$

To compute this derivative, let the denominator of the above limit be denoted $g(z)$. Thus,

$$\left(\frac{1}{g(z)} \right)'' = \left(-\frac{g'(z)}{g^2(z)} \right)'$$

$$= \frac{-g''(z)g^2(z) + g'(z) \cdot 2g(z)g'(z)}{g^4(z)}$$

From above definitions and the series for $g(z)$, the values for $g(0)$, $g'(0)$, and $g''(0)$ may be computed, resulting in the final solution

$$\frac{1}{2} \cdot \frac{\dfrac{1}{3} \cdot 1^2 - 0}{1^4} = \frac{1}{6}.$$

CHAPTER 9

1. $\pi/3$

2. $\pi/6$

4. $\pi\sqrt{3}/6$

5. $\pi/12$

7. $\displaystyle\int_{C_r} \frac{ze^{iz}}{z^2+1}\,dz = \frac{\pi i}{e}$;

$$\left| \int_{BCA} \frac{ze^{iz}}{z^2+1}\,dz \right| \le \int_{BCA} \frac{|z||e^{iz}|}{|z^2+1|}\,|dz|$$

$$\le \int_0^\pi \frac{re^{-y}}{r^2-1}\left|\frac{dz}{dt}\right|\,dt = \frac{r^2}{r^2-1}\int_0^\pi e^{-r\sin t}\,dt$$

$$\le \frac{r^2}{r^2-1}\int_0^{\pi/2} e^{-rat}\,dt = \frac{r^2}{r^2-1}\left(-\frac{1}{ar}e^{-rat} \right)\Big|_0^{\pi/2}$$

$$= \frac{r/a}{r^2-1}(1-e^{-r}) < \frac{r/a}{r^2-1} \to 0 \text{ as } r \to \infty$$

(see Example 3 for details).

9. $0 = \displaystyle\int_{C_r} e^{iz^2}\,dz = \int_{AB} e^{iz^2}\,dz$

$$+ \int_{BD} e^{iz^2}\,dz + \int_{DA} e^{iz^2}\,dz\,;$$

$$\left| \int_{BD} e^{iz^2}\,dz \right| \le \int_{BD} |e^{iz^2}|\,|dz| = \int_0^{\pi/4} |e^{iz^2}|\,|rie^{it}|\,dt$$

$$= r\int_0^{\pi/4} |e^{ir^2(\cos 2t + i\sin 2t)}|\,dt = r\int_0^{\pi/4} e^{-r^2\sin 2t}\,dt$$

$$\le r\int_0^{\pi/4} e^{-r^2 at}\,dt = \frac{1}{2ar}(1-e^{-2r^2}) \to 0$$

$(a = 4/\pi)$;

Segment AD is $z = x(1+i)$, $0 \le x \le r \Rightarrow$

$$\int_{AD} e^{iz^2}\,dz = \int_0^r e^{ix^2(1+i)^2}(1+i)\,dx$$

$$= (1+i)\int_0^r e^{-2x^2}\,dx = (1+i)\int_0^{r\sqrt{2}} e^{-u^2} \cdot \frac{du}{\sqrt{2}}$$

$$\to \frac{1+i}{\sqrt{2}} \cdot \frac{\sqrt{\pi}}{2} = (1+i)\frac{\sqrt{2\pi}}{4} \text{ as } r \to \infty$$

$$0 = \lim_{r \to \infty} \int_{C_r} e^{iz^2}\,dz \quad 0$$

$$= \lim_{r \to \infty} \int_{AB} e^{iz^2}\,dz + 0 - (1+i)\frac{\sqrt{2\pi}}{4}$$

Therefore, with $z = x$ (real),

$$\lim_{x \to \infty} \int_{AB} (\cos x^2 + i\sin x^2)\,dx = (1+i)\frac{\sqrt{2\pi}}{4}$$

$$\int_0^\infty \cos x^2\,dx + i\int_0^\infty \sin x^2\,dx = \frac{\sqrt{2\pi}}{4} + i\frac{\sqrt{2\pi}}{4}$$

11. (b) $w = cz$ where $c = 3 + 4i \Rightarrow \theta = \tan^{-1}4/3 \approx 53.1°$

13. $w = z' = \dfrac{1}{2+iy} \Rightarrow |z' - \frac{1}{4}| = \left| \dfrac{1}{2+iy} - \dfrac{1}{4} \right|$

$$= \left| \frac{2-iy}{8+4iy} \right| = \sqrt{\frac{4+y^2}{64+16y^2}} = \frac{1}{4}$$

14. $C = a/c$, $D = (bc - ad)/c$.

15. $C = 3$, $D = -1$ in Problem 14. **(a)** $w_1 = z + 1$, $w_2 = 1/w_1$ and $w = -w_2 + 3$; **(b)** $w_1 = z + 1$ so circle is translated to $|z + 1| = 1$; $w_2 = (z + 1)^{-1}$ is inversion reflected in y-axis, so circle $|z + 1| = 1$ (passing through origin) maps to vertical line $z = -\frac{1}{2}$, which is then mapped to $z = -(-\frac{1}{2}) + 3 = \frac{7}{2}$ (vertical line); **(c)** Horizontal line $y = -\frac{1}{2}$ maps to itself under w_1, then is inverted to circle $|z - i| = 1$ under w_2, and finally reflected through origin to $|z - i| = 1$ and translated 3 units to right to circle $|z - 3 + i| = 1$.

20. The horizontal line $(-\pi/2 < x < \pi/2, y = b > 0)$ maps to the ellipse $(u/\cosh b)^2 + (v/\sinh b)^2 = 1$, $v > 0$, with $u = \sin x \cosh b$, so if P and Q are distinct points in the xy-plane and lie on the same horizontal line, their images P' and Q' are distinct points in the uv-plane. If P and Q lie on different lines $y = b < b'$, the u and v intercepts of the ellipse for $y = b$ lie inside the ellipse for $y = b'$ (since $\cosh b < \cosh b'$, $\sinh b < \sinh b'$ for b, $b' > 0$). Hence $P' \neq Q'$ (similarly for b, $b' < 0$). This leaves points on the vertical lines $x = \pm\pi/2$ to examine. First, if $P = (\pi/2, b)$ for $b > 0$ and $Q =$ any other point, then $P' = (\cosh b, 0)$ and either $Q' = (\cosh b', 0)$ or $Q' = (u, c)$ for $c \neq 0 \Rightarrow P' \neq Q'$. The proof for $b < 0$ is similar. Finally, if $b = 0$ the x-axis $(-\pi/2 \leq x \leq \pi/2)$ maps one-to-one to the u-axis $(-1 \leq u \leq 1)$. (Remaining points on the u-axis are the images of unique points not on the x-axis.)

21. (b) $(e^x\cos y)^2 = (e^a\cos b)^2$ and $(e^x\sin y)^2 = (e^a\sin b)^2$ $\Rightarrow e^{2x} = e^{2a}$ and $x = a$. Hence $\cos y = \cos b$ and $\sin y = \sin b$. The graphs of sine and cosine on the interval $(-\pi, \pi]$ show that $y = b$. [On this interval, $\cos y = \cos b$ implies $y = \pm b$ and, unless $y = b$, $\sin y = \sin b$ implies $\sin y = 0 = \sin b$; thus either $y = 0$ and $b = \pi$, or $y = \pi$ and $b = 0$, contradicting $\cos y = \cos b$. Therefore, $y = b$.]

24. (a) $w_1 = 2$, $w_2 = -1$, $w_3 = \frac{1}{2}(\sqrt{5} - 1)$, $w_4 = -\frac{1}{2}(\sqrt{5} + 1)$. For $1 \leq k \leq 4$, $g'(w_k) = 16$, 4, -4, or $-4 + 8\sqrt{5}$; **(b)** $w_3 = \frac{1}{2}(\sqrt{5} - 1)$.

25. (b) Since $c_{n+1} = c_n^2 + c$ for all n, taking the limit as $n \to \infty$ yields $a = a^2 + c$. Solving this quadratic in a produces the roots as in **(9.8)**.

26. Suppose $c \in [-2, \frac{1}{4}]$, then, for $c \geq 0$, $c_2 = c^2 + c < \frac{1}{16} + \frac{1}{4} = \frac{5}{16} < \frac{1}{2} \Rightarrow 0 \leq c_2 < \frac{1}{2}$. By mathematical induction, if $0 \leq c_n < \frac{1}{2}$, $c_{n+1} = c_n^2 + c < \frac{1}{4} + \frac{1}{4} = \frac{1}{2}$ and $0 \leq c_{n+1} < \frac{1}{2}$. Therefore $\{c_n\}$ is bounded and $c \in M$. For $c < 0$ then $-2 \leq c < 0 \Rightarrow c = -2 + \varepsilon$ for some ε such that $0 \leq \varepsilon < 2 \Rightarrow c_2 = c^2 + c = (-2 + \varepsilon)^2 - 2 + \varepsilon = 2 - 3\varepsilon + \varepsilon^2$. Using simple inequalities, $-4 < -3\varepsilon + \varepsilon^2 < -\varepsilon$ so that $-2 < c_2 < 2 - \varepsilon$. Inductively, assume $-2 < c_n < 2 - \varepsilon$ has been proved. Then $c_{n+1} = c_n^2 + c < (2 - \varepsilon)^2 - 2 + \varepsilon = 2 - 3\varepsilon + \varepsilon^2$, and as before, $-2 < c_{n+1} < 2 - \varepsilon$. Hence $\{c_n\}$ is bounded and $c \in M$. Conversely, suppose $c \in M$ and c is real. By **(9.9)**, $c \geq -2$; it remains to prove that $c \leq \frac{1}{4}$. Suppose instead that $c > \frac{1}{4}$. Then $\{c_n\}$ is increasing: $c_{n+1} = c_n^2 + c > c_n^2 + \frac{1}{4} - c_n + c_n = (c_n - \frac{1}{2})^2 + c_n \geq c_n$. Since $\{c_n\}$ is bounded, $\lim c_n$ exists and converges to $\frac{1}{2}(1 + \sqrt{1 - 4c})$ by **(9.8)**, which is real $\Rightarrow 1 - 4c \geq 0$, a contradiction. Therefore, $c \leq \frac{1}{4} \Rightarrow c$ lies on the interval $[-2, \frac{1}{4}]$.

27. By definition,
$$\ln z = \ln\sqrt{x^2 + y^2} + i(\tan^{-1} y/x + k\pi)$$
where k is an integer depending on x and y, which is analytic. Hence $\tan^{-1} y/x$ is a harmonic conjugate.

35. (a) $\left(1 - \dfrac{1}{2}\right)\left(1 + \dfrac{1}{3}\right)\left(1 - \dfrac{1}{4}\right)\left(1 + \dfrac{1}{5}\right)\left(1 - \dfrac{1}{6}\right)\cdots$
$$= \left(\dfrac{1}{2}\right)\left(\dfrac{4}{3}\right)\left(\dfrac{3}{4}\right)\left(\dfrac{6}{5}\right)\left(\dfrac{5}{6}\right)\cdots = \dfrac{1}{2}$$

(b) The partial products are
$$p_n = \left(1 - \dfrac{1}{2}\right)\left(1 + \dfrac{1}{3}\right)\left(1 - \dfrac{1}{4}\right)\left(1 + \dfrac{1}{5}\right)\left(\dfrac{5}{6}\right)\cdots$$
$$\cdot\left(1 + \dfrac{1}{2n-1}\right)\left(1 - \dfrac{1}{2n}\right)\delta_n$$

where $\delta_n = 1$ or $1 - 1/(2n + 1)$ depending on whether n is even or odd. This product equals
$$p_n = \dfrac{1}{2}\left(\dfrac{4}{3}\right)\left(\dfrac{3}{4}\right)\left(\dfrac{6}{5}\right)\left(\dfrac{5}{6}\right)\cdots\left(\dfrac{2n}{2n-1}\right)$$
$$\cdot\left(\dfrac{2n-1}{2n}\right)\delta_n = \dfrac{1}{2}\delta_n$$

and $\lim p_n = \frac{1}{2} \cdot \lim \delta_n = \frac{1}{2}$ as $n \to \infty$. By definition, the infinite product converges to $\frac{1}{2}$).

39. $\pi/4$

47. $\displaystyle\int_0^z \tan w\, dw = \sum_{n=1}^{\infty}\int_0^z \dfrac{8w}{(2n-1)^2\pi^2 - 4w^2}\, dw;$

$\displaystyle -\ln(\cos w)\Big|_0^z = -\sum_{n=1}^{\infty}\ln[(2n-1)^2\pi^2 - 4w^2]\Big|_0^z;$

$\displaystyle \ln(\cos z) = \sum_{n=1}^{\infty}\ln\left(\dfrac{(2n-1)^2\pi^2 - 4z^2}{(2n-1)^2\pi^2 - 4\cdot 0^2}\right);$

$\displaystyle \ln(\cos z) = \sum_{n=1}^{\infty}\ln\left(1 - \dfrac{4z^2}{(2n-1)^2\pi^2}\right).$

APPENDIX B:

BIBLIOGRAPHY

Barnsley, M.F., 1988. *Fractals Everywhere*. New York: Academic Press.

Burton, D.M., 1991. *The History of Mathematics: An Introduction, 2nd Ed.* Dubuque, Iowa: Wm. C. Brown.

Churchill, R.V. and Brown, J.W., 1990. *Complex Variables and Applications, 5th Ed.* New York: McGraw-Hill.

Eves, H., 1966. *Functions of a Complex Variable*. Boston: Prindle, Weber and Schmidt.

Hardy, G.H. and Littlewood, J.E. Abel's theorem and its converse. *Proc. London Math. Soc.* (2), 18 (1918), 205–235.

Spiegel, M.R. and Lipschutz, S., 2009. *Complex Variables, 2nd Ed (Schaum's Outline Series)*. New York: McGraw-Hill.

Titchmarsh, E.C., 1975. *The Theory of Functions, 2nd Ed.*, Oxford University Press.

INDEX